T0350926

Advances in Biomedical and Biomimetic Materials

Advances in Biomedical and Biomimetic Materials

Ceramic Transactions, Volume 206

*A Collection of Papers Presented at the
2008 Materials Science and Technology
Conference (MS&T08)
October 5–9, 2008
Pittsburgh, Pennsylvania*

Edited by

R. J. Narayan
P. N. Kumta
W. R. Wagner

A John Wiley & Sons, Inc., Publication

Published by John Wiley & Sons, Inc., Hoboken, New Jersey.
Published simultaneously in Canada.

For general information on our other products and services or for technical support, please contact our Customer Care Department within the United States at (800) 762-2974, outside the United States at (317) 572-3993 or fax (317) 572-4002.

Wiley also publishes its books in a variety of electronic formats. Some content that appears in print may not be available in electronic format. For information about Wiley products, visit our web site at www.wiley.com.

Library of Congress Cataloging-in-Publication Data is available.

ISBN 978-0-470-40847-6

Printed in the United States of America.

10 9 8 7 6 5 4 3 2 1

Contents

MATERIALS FOR DRUG DELIVERY

Preface

This issue contains the proceedings of the "Advances in Biomedical and Biomimetic Materials" symposium, which was held on October 5–9, 2008 at the David L. Lawrence Convention Center in Pittsburgh, Pennsylvania, USA. The development of materials for dental and medical applications is a rapidly developing focus of activity in materials science and engineering. Novel processing, characterization, and modeling techniques continue to be developed that will provide enhanced diagnosis and treatment of medical conditions. Presentations were given on recent developments in biomedical and biomimetic materials, including scaffolds for tissue engineering; bioceramics; biomimetic materials; surface modification of biomaterials; metallic implant materials; nanoparticles for medical diagnosis and treatment; as well as novel materials for drug delivery and biosensing. This symposium enabled discussion among the many groups involved in the development and use of biomaterials, including materials researchers, medical device manufacturers, and clinicians.

We would like to thank the staff at The American Ceramic Society for making this proceedings volume possible. We also give thanks to the authors, participants, and reviewers of the proceedings issue. We hope that this issue becomes a significant resource in the areas of biomedical materials research and biomimetic materials research that not only contributes to the overall advancement of these fields but also signifies the growing roles of The American Ceramic Society, ASM International (The Materials Information Society), The Minerals, Metals & Materials Society, and the Association for Iron and Steel Technology in these rapidly developing areas.

R. J. NARAYAN, *University of North Carolina and North Carolina State University*

P. N. KUMTA, *University of Pittsburgh*

W. R. WAGNER, *University of Pittsburgh*

Bioceramics

BIOTRIBOLOGICAL CHARACTERIZATION OF THE BILAYER SYSTEM: HA/ZRO$_2$ ON 316LSS

B. Bermúdez-Reyes,[1,2] I. Espitia-Cabrera,[3] J. Zárate-Medina,[1] M. A. L. Hernández-Rodríguez,[4] J. A. Ortega-Saenz,[4] F. J. Espinoza-Beltrán,[2] M. E. Contreras-García[1]

[1] Instituto de Investigaciones Metalúrgicas de la Universidad Michoacana de San Nicolás de Hidalgo. Edificio U, Cuidad Universitaria. Av. Francisco J. Mújica s/n. Colonia Felicitas del Río. Morelia, Michoacán, México.

[2] Centro de Investigación y Estudios Avanzados del I. P. N. Unidad Querétaro. Libramiento Norponiente # 2000. Fraccionamiento Real de Juriquilla. Santiago de Querétaro, Querétaro, México.

[3] Facultad de Ingeniería Química de la Universidad Michoacana de San Nicolás de Hidalgo. Edificio M, Cuidad Universitaria. Av. Francisco J. Mújica s/n. Colonia Felicitas del Río. Morelia, Michoacán, México.

[4] Facultad de Ingeniería Mecánica y Eléctrica, Universidad Autónoma de Nuevo León. Av. Universidad s/n. San Nicolás de los Garza, Nuevo León, México.

ABSTRACT

Orthopedic prostheses have to be highly resistant to wear and physiological corrosion. The bilayer system HA/ZrO2/316LSS is proposed in this work as a coating that fulfills these requirements, because it a presents very good physiological corrosion resistance. The hydroxyapatite coating was deposited by screen printing on 316LSS previously covered by electrophoresis with a zirconia thin film and thermally treated at 650 °C for 5 min. The biotribological characterization was carried out on 316LSS hip heads and acetabular cups in Hip Simulator FIME II equipment, using bovine fetal serum (BFS) as a lubricant. Scanning electron microscopy images were obtained at the beginning and at the end of the test in order to observe the wear produced on the samples. EDS chemical analysis was also obtained. From the obtained results, the role of the hydroxyapatite as a solid lubricant was elucidated and it was concluded that the bilayer system actually works efficiently to protect the 316LSS prosthesis under extreme working conditions.

INTRODUCTION

New materials have been designed for specific implants and organs. In the last century, millions of people have had the need to use dental implants, bone substitutes (prostheses and refills) and hybrid organs (digestive apparatus parts) [1].

The human body is an engineering work; however, wear and degradation are part of the useful life, and, as with any machine, after failure some parts have to be substituted by implants or prostheses. Materials Science Scientifics have recently been trying to provide a solution to this specific requirement. In this application all the materials science areas are involved because the human body is formed of ceramics (hydroxyapatite, fluoroapatite), metals (Fe, Cr, Ca), polymers (keratin), composites (myosin and actin) and organic materials (albumin, glucose, cholesterol) [2]. Moreover, the human body transports substances and nutrients necessary for chemical stability using a fluid composed mainly of chlorides, called blood plasma. The blood plasma has a pH of 7.4, indicating that there is a lightly basic environment. However, it is highly corrosive and degrade

3

316LSS, Co-Cr and titanium alloys. Besides the problem of degradation by corrosion, the implants are also subject to wear, which can be attributed to three different causes: natural wear, chronic-degenerative diseases and lack of lubrication [3]. So the prosthesis must present mechanical stability, wear and corrosion resistance and biocompatible properties [4].

Biotribology is an area of tribology with applications in biomechanics, biomaterials, orthopedic and biologic systems [5]. Thus biotribology studies friction, wear and lubrication in diarthrosis systems like the hip joint [6].

The biotribological test has been designed to test whether materials reach the standard wear resistance in a physiological medium, simulating the movement conditions of the prosthesis in the body's service.

An osseous joint is a low friction complex mechanism that permits movement and transmission of load from bone to bone. The bones end in joints covered by articular cartilage tissue of about 2 mm thickness and lubricated by synovial liquid, in amounts of approximately 0.5–2 ml [7, 8].

The hip joint is the joint that demands the greatest wear resistance. For this reason, biotribological equipment is designed to simulate femoral heads. This joint presents simultaneous static and dynamic momentums composed of six movements in three sectional planes during the walk cycle. These movements are flexion-extension (FE), abduction-adduction (AA) and internal-external rotation (IE). All these movements are presented simultaneously within a period of approximately 1.1 seconds, with a charge-discharge cycle called Paul's cycle that simulates the charge produced during the heel lean on the floor, the oscillation and the other heel lean [9]. The biomechanical simulators have been designed with these walk cycles and Paul's cycles.

In the biotribological test, the physiological condition simulation as well as the biomechanical simulation must be controlled. This test presents extreme and accelerated conditions of load and movement. The results must confirm that the geometry and material design are adequate under these extreme operating conditions [10].

The biotribological behavior of the HA/ZrO$_2$/316LSS bilayer system proposed as a biomaterial is presented and analyzed in this work.

EXPERIMENTAL PROCEDURE

Substrate Preparation

The 316LSS femoral heads were schemed in a HAUS wheel; model HAA5, with ¾ of the sphere, 30 mm in diameter, joined to an offshoot 50 mm in length. The 316LSS acetabular cups were also machined in a Mazak wheel, model NEXUS250 II, with a 30 mm diameter and joined to an offshoot 30 mm in length. The substrate used was 316LSS because it is the material commonly used for prostheses in the Public Assistance Health System in Mexico. This alloy has a chromite thin film on the surface, and this oxide plays an important role in the anchorage of the bilayer coating as was described in a previous work [11]. The femoral heads and acetabular cups were polished with SiC sandpaper from 400 to 4000 mesh and with alumina of three different sizes: 1

μm, 0.3 μm and 0.05 μm. The samples were mirror polished; a final roughness of 0.06 μm was measured in both cases by a Taylor Hobson LTD profilometer.

ZrO$_2$ coating application

The 316LSS femoral heads and acetabular cups substrate was coated with a ZrO$_2$ film using the electrosynthesis deposition method (EDP). The deposition solution used in this process was 0.005M ZrOCl$_2$ (Aldrich)/water, and the electrodeposition was carried out by applying a bias potential of 9 mV with a deposition time of 90 sec. The deposited films were dried at 100 °C for 30 min in order to attain the complete elimination of the HCl formed during the ZrOCl$_2$ hydrolysis; the description of the electrodeposition conditions can be found in a previous work [11].

HA coating application

The hydroxyapatite (HA) coating was made by using a screen printing technique on the ZrO$_2$/316LSS. The HA paste was elaborated with HA (Alfa-Aesar) and propylenglycol (Baker) with a rate of 7:3. The paste was applied through a polymeric mesh with 120 threads/cm^2. The propylenglycol was evaporated by thermal treatment at 200 °C for 10 min. The complete bilayer HA/ZrO$_2$/316LSS system was thermally treated at 650 °C for 5 min. The deposition screen printing method and conditions were also described in a previous work [11]. The interface formed with the thermal treatment between HA and ZrO$_2$ coatings was analyzed by scratching the HA coating with a stainless steel spatula. Figure 1 shows photographs of the femoral heads and acetabular cups coated with the bilayer system and the interface. Hence the femoral heads and acetabular cups that were tested were 316LSS, ZrO$_2$/316LSS, HA/ZrO$_2$/316LSS and the HA/ZrO$_2$ interface.

Figure 1.- Biotribological test joins.

All samples were tested in the FIME II biotribometer [12] in anatomical position and the AA, FE and IE movements were simulated. Every station was lubricated with 62.5 ml of fetal bovine serum diluted in 187.5 ml distilled water as recommended by Tiina Ahlroos [8]. After each step of 4 x 10^3 cycles the serum was changed, the articulations were washed and the weight of each femoral head and acetabular cup was registered. The tests continued until they reached 2 x 10^4

cycles, which is equivalent to 4×10^4 steps with a load of 3 KN, which is equivalent to the weight of a mass of 300 kg.

The joints morphology was analyzed before and after the test in two JEOL scanning electron microscopes, JSM-6490LV and 5910LV models, at low vacuum with backscattering electrons. Chemical analysis was carried out in a Phillips model XL30ESEM scanning electron microscope at high vacuum.

RESULTS AND DISCUSSION

In general, the expected service of an orthopedic prosthesis and in particular the most modern hip prosthesis is from 12 to 15 years. This is a long period; however, it can be reduced due to physiologic attack and wear on the prosthesis [13].

In figure 2, the movements and loads applied to the joints during the biotribological test are shown in accordance with the ISO 14242 norm [14].

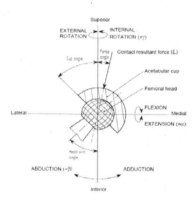

Figure 2.- Mobility and loads on hip join [14].

Figure 3 shows the SEM images of 316LSS substrates. Figure 3a shows the naked 316LSS substrate surface before the biotribological test: it presents some porosity. Figure 3b shows a low magnification (100×) SEM image of the 316LSS substrate after 20000 cycles of the biotribological test. The surface presents damage in several places, consisting of deep disordered furrows and deep holes. Images of different zones at different magnifications are presented in figures 3c (250×) and 3d (2000×), showing the damage produced by wear and fatigue of the substrate during the test. It is evident that the 316LSS substrate suffered loss of material during the test.

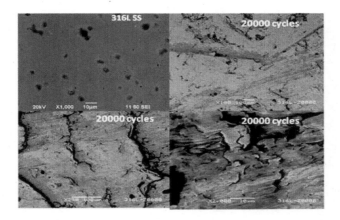

Figure 3. 316L SS SEM images a) before Biotribologycal test and after of 20000 cycles b) 100X, c) 250X and d) 2000X.

The different and darker gray zone observed and indicated in the backscattered electron-micrograph of figure 3b indicates the presence of a deposit of a different composition on the furrow, which may be a protein deposit in accordance with the reports by Chevalier et al. [19] and Caton [20]. The protein deposit originated from the bovine fetal serum used as lubricant in the biotribological test. Similar wear and fatigue damage was reported by Nakajima et al. on 304 SS in a fatigue test under physiological conditions; they detected loss of material and holes and marks, which were attributed to fatigue [15].

Figure 4 shows SEM images of the ZrO$_2$/316LSS system. The micrograph in figure 4a at 1000× presents a smooth and uniform zirconia film surface. The micrograph at 100×, after 20000 cycles of the biotribological test, is shown in figure 4b; it is evident that the damage to the sample caused by wear and fatigue provoked deep furrows. The micrograph at a higher magnification (250×) in figure 4c presents a zone with a big hole. At a higher magnification (2000×) the backscattered micrograph presents the damage to the sample with furrows that do not contain any deposits of a different composition. It is evident that there is no zirconia deposit on the surface. De Aza et al. reported that zirconia does not have the property of fixing proteins on the surface. Proteins are usually fixed on the surface due to the production of localized corrosion points; however, due to its high corrosion resistance, the tetragonal zirconia does not permit proteins to be fixed on it. The opposite case is presented by the monoclinic phase of zirconia, which is corrosion susceptible [18]. Patel and Spector obtained similar results in the analysis of ZrO$_2$/UHMWPE friction pairs; they found proteins adhered to the UHMWPE surface but not to the zirconia surface [19].

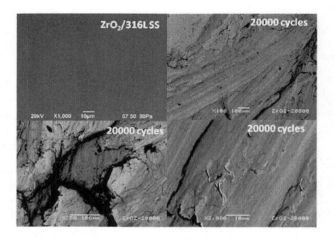

Figure 4. ZrO$_2$/ 316L SS SEM images a) before Biotribologycal test and after of 20000 cycles
b) 100X, c) 250X and d) 2000X.

Figure 5 shows SEM images of the HA/ZrO$_2$/316LSS system. In figure 5a, the porosity and roughness of the surface before the biotribological test are shown. In figure 5b, after 20×10^3 cycles, it is observed that the bilayer coating has disappeared and that circular furrows containing material of different composition are now present. At an amplification of 2000×, protein deposits (figures 5c and 5d) can be observed in these furrows. The proteins deposits are remains of the fetal bovine serum used as lubricant throughout the test.

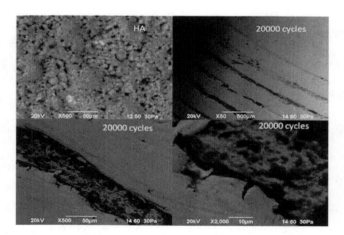

Figure 5. HA/ZrO$_2$/ 316L SS SEM images a) before biotribological test and after of 20000
cycles b) 50X, c) 500X and d) 2000X.

To determine that the remains in the furrows are proteins, EDS microanalyses were performed. These measurements confirmed that the remains in the furrows predominantly contained phosphorous, which corresponds to phosphorus based proteins from the fetal bovine serum. Also, Ca and O were detected, which may possibly be the remains of the HA coating, and some elements (Fe, Ni and Cr) of 316LSS were detected too (figure 6). To verify the loss of the HA and ZrO₂ coatings, EDS was also carried out outside the furrows, and only the 316LSS elements were detected (figure 7).

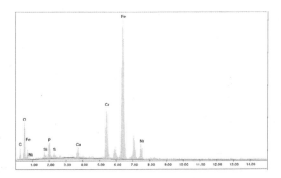

Figure 6. EDS microanalysis from furrows remain.

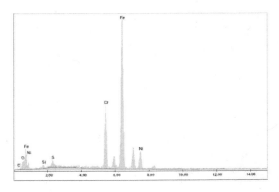

Figure 7. EDS microanalysis from out furrows.

Hermanson and Söremark detected that in HA-ZrO$_2$ bifunctional composites, the mechanical properties were found to be good after evaluation of their properties using NaCl solution as a lubricant at 37°C. They observed an important effect of slip due to HA. However, the HA remains are not toxic, because the HA is important to the general mineral organism stability during the bone's natural wear [16].

On the other hand, Wang et al. determined that a HA/ZrO$_2$ partially stabilized system has the property of attracting and fixing proteins, due to which the system presents an efficient marginal lubrication, and this lubricating property was noticeable during the wear test [17].

It should be pointed out that in this work; the hip joint lubrication is of the marginal type. This means that the tested hip joint has with the minimum lubricant necessary between the femoral head and the acetabular cup.

Figure 8 shows SEM images of the HA/ZrO$_2$ interface. Figure 8a shows a smooth and uniform surface of the HA/ZrO$_2$ interface before the biotribological test. After 20000 cycles, a protein layer adhered to the surface is observed in figure 8b. Figures 8c and 8d, with more amplification, show the morphology inside and around furrows of other zones, with a lot of damage and protein deposits. It is observed that the sample was damaged by severe friction (figure 8c). Figure 8d shows the interior of furrows and there are observed remains of proteins. This is confirmed by EDS analysis, which detected phosphorus, calcium and the 316LSS components (figure 9).

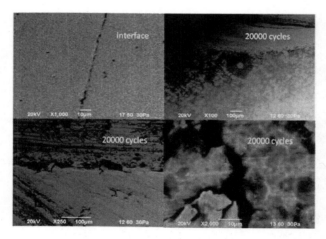

Figure 8.- HA/ZrO$_2$ interface SEM images a) before biotribological test and after of 20000 cycles b) 100X, c) 250X and d) 2000X.

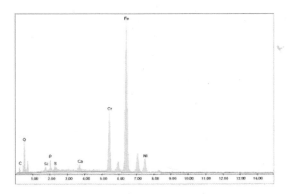

Figure 9.- EDS microanalysis from surface interface

The proteins adherence, according to Chevalier et al., is due to the martensitic tetragonal to monoclinic zirconia transformation due to friction. They observed this transformation during tests using water and fetal bovine serum as lubricant [20]. Caton also detected the martensitic tetragonal to monoclinic transformation and the proteins' adherence on a zirconia surface. He determined that the adherence of proteins on the surface generates incipient corrosion points [21]. On the other hand, Paff and Willmann determined that zirconia hydrothermal decomposition *in vivo* is delayed by approximately 10 years. However, when monoclinic zirconia appears and is heterogeneously distributed, it produces stress points that generate decrements in the mechanical properties, hence the pure zirconia prosthesis presents imminent failure *in vivo* [22]. Moreover, Piconni et al. detected fractures in the femoral heads, originated by the zirconia hydrothermal transformation. They also observed a proteins cumulus near the fractures and pitting corrosion on the zirconia surface [23].

The surface evolution of the sample throughout the biotribological test and the analysis of the loss of weight of the wearing pieces give important information about the detached material that could go into the bloodstream from the prosthesis inside a human body.

The initial weights of the samples before the biotribological test are reported in Table 1. Following the test, the weight lost by each tested sample is reported in Table 2. From this data, figure 10 shows the graphs of weight loss versus number of cycles of the 316LSS, ZrO$_2$/316LSS, HA/ZrO$_2$/ 316LSS and HA/ZrO$_2$ interface femoral heads and acetabular cups. The complete system lost more weight than the HA/ZrO$_2$ interface, but the interface presents more damage than the complete system. This is due to the HA coating thickness, which apparently has lower wear resistance than the HA/ZrO$_2$ interface, so the HA is detached from the deposit in the first cycles and then behaves like solid lubricant. The ZrO$_2$/316LSS system presented a very similar weight loss to the total system until 12000 cycles, after which an abrupt fall in the weight of the acetabular cups was observed. For the femoral heads, the system presented an initial weight loss from the first cycles, but this loss is less than the loss of the naked 316LSS, which indicates that the zirconia film is actually protecting the 316LSS surface. Then the weight loss stabilized from 8000 to 12000 cycles, after which the weight loss accelerated. This different behavior between the case of the acetabular cups and that of the femoral heads could be due to the fact that in the test the acetabular cup is fixed to the upper offshoot and the femoral head is in constant movement. It can be concluded that the zirconia film protects the 316LSS until 12000 cycles, after which the zirconia film was destroyed in the test.

Table 1. 316LSS and Coatings total weight before the biotribological test

Material	Femoral ball weight	Acetabular cup weight
316L SS	128.960 g	161.385 g
ZrO$_2$/316L SS	129.433 g	161.617 g
HA/ZrO$_2$/316LSS	129.208 g	161.657 g
HA/ZrO$_2$ Interface	128.200 g	161.656 g

Table 2. Total weight lost after the test.

Material	Total weight lost of the femoral ball	Total weight lost of the acetabular cup
316L SS	2.201 g	1.014 g
ZrO$_2$/316L SS	2.697 g	1.163 g
HA/ZrO$_2$/316LSS	0.074 g	0.046 g
HA/ZrO$_2$ Interface	0.021 g	0.013 g

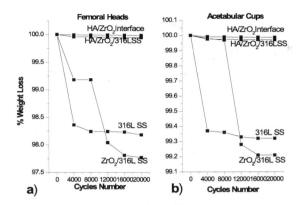

Figure 10.- Plots of percentage of weight lost versus number of cycles for a) femoral heads and b) acetabular cups.

Similar behavior was reported by Affatato et al., who analyzed ZrO$_2$–ZrO$_2$, ZrO$_2$–Al$_2$O$_3$ and HA/ZrO$_2$–HA/ZrO$_2$ hip joints in physiological conditions with fetal bovine serum as a lubricant. In this work, they determined that ZrO$_2$–ZrO$_2$ pairs presented good wear resistance but they also detected zirconia hydrothermal transformation. In ZrO$_2$–Al$_2$O$_3$ pairs, they detected a reduction of the fracture incidence, and in the HA/ZrO$_2$ –HA/ZrO$_2$ pairs they found significant improvements in the wear resistance and the material did not lose biocompatibility properties [24]. Baroud and Willmann also determined that HA coatings tend to become detached, but the debris produced by the wear are not toxic, and for this reason the coating did not loose the biocompatibility property [25].

Lawn determined that trilayer structures (ceramic/ceramic/substrate) show good acceptance in the biomechanics area, especially in dental devices and in orthopedic prostheses. This is because these structures perform like core-armor. The ceramic/ceramic part acts like armor and this is the system's functional part. The ceramic/substrate part acts like a nucleus, protecting the substrate from physiologic attack, and for these reasons the materials design should be very meticulous [26]. The HA/ZrO$_2$/316LSS system studied in this work actually acts in the way described by Brian R. Lawn.

It is worthwhile to depict the marks detected on the femoral heads and the acetabular cups surfaces, which are due to the walk and Paul's cycles. These marks coincide well with mark patterns described by Colonius and Saikko in 2002 [29, 30]. Figure 11 shows the patterns formed during the biotribological test and the comparison of them with the Paul's cycles (figure 11).

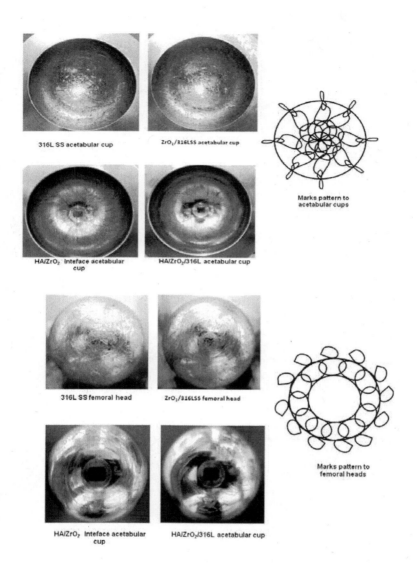

Figure 11 marks produced during biotribological test in 316L SS, ZrO$_2$/316L SS, HA/ZrO$_2$/316L SS and HA/ZrO$_2$ interface femoral balls and acetabular cups, compared with the marks patterns

Reinisch, Judmann and Pausschitz, based on the ISO 14242, determined that 5,000,000 cycles are equivalent to 5 years *in vivo* service [27]. With this data, it was determined that 20000 cycles of the system are equivalent to 7.3 days in service. Bergman determined that a person walks at 6 km/hr on average [28]. Hence, it is possible to determine that in 7.3 days a person with a hip prosthesis would walk 1051.2 km. This paper analyzes the behavior of the HA/ZrO$_2$/316LSS system during 20000 cycles of the biotribological test under extreme wear conditions in order to elucidate expectations of the proposed system in its application as a coating for hip joints.

CONCLUSIONS

A biocompatible HA/ZrO$_2$ bilayer on 316LSS substrates was tested in a biotribometer machine. It was determined that this bilayer coating on steel substrates has low wear resistance when working at the extreme conditions of the hip biotribological tests. These results evidence the convenience of developing new, softer biotribological tests for evaluating this kind of biocompatible coating.

The biotribological tests of the hip-joint samples reported in this work were performed under similar conditions described in the test of ISO 14242. However, the number of cycles was limited at 2×10^4, which is significantly lower than the 3 or 5×10^6 cycles commonly used for this ISO test. It indicates that the proposed HA/ZrO$_2$/316LSS system does not show good behavior under the extreme conditions tested for hip prosthesis applications. However, this work made it possible to evaluate the behavior of the HA/ZrO$_2$ bilayer on 316LSS substrates under very demanding conditions in physiological media. According to the results obtained, the HA/ZrO$_2$ bilayer on 316LSS substrates could be adequate as a biocompatible and wear protective coating on prostheses with low tribological demands, such as the buccal prosthesis, and can also be recommended for scaffold applications.

ACKNOWLEDGMENTS

We thank Miguel Ángel Quiñones Salinas, María Eugenia Ávila Rodríguez, Ruth Viridiana Martínez López and Antonio Olivares from FIME-UANL for their valuable help, and Diego Lozano from NEMAK for his disposition to roughness measurements; Héctor Damián Orozco Hernández from IIM-UMSNH for his valuable aid in producing zirconia coatings on 316LSS; Javier Israel Alvarado from COMIMSA for his help in the 316LSS articulation preparation; Dra. Lourdes Mondragón Sánchez from ITM for obtaining SEM images and Ing. Eleazar Urbina from CINVESTAV-Qro for EDS analysis.

REFERENCES

[1] Kevin E. Healy, Alereza Rezania and Ranee A. Stile. Desing Biomaterials to Direct Biological Responses. Annals New York Academy Sciences, 24-25 (2006).

[2] Atlas Ilustrado de Anatomía. Susaeta Ediciones. Madrid España (2002).

[3] Jessica N. Albino Soto, Nitza Correa Cora and Jessenia Filiberty Irzaberry. Mechanics of biomaterials: Orthopaedics. University of Puerto Rico. May 2005.

[4] Buddy D. Ratner, Allan S. Hofmann, Frederick J. Shoen, Jack Lemons. Biomaterials Science. An introduction to Materials in Medicine. Elsevier academic Press (2004).

[5] Thomas P. Schmalzreid and John J. Callaghan. Wear in total Hip and Knee Replacements. The Journal of Bone and Joint Sugery. Vol. 81-A, No. 1, January 1999.

[6] Nils A. Steika Jr. A Comparison of the Wear Resistance of Normal, Degenerate, and Repaired Human Articular Cartilage. Thesis submitted to the faculty of Virginia Polytechnic Institute and State University. October 29, 2004.

[7] Atlas Ilustrado de Anatomía. Susaeta Ediciones. Madrid España (2002).

[8] Tiina Ahlroos. Effect of lubricant on the wear of prosthetic joint material. Acta Polytecnica Scandinavica (2001).

[9] Javier Alonso Ortega Saenz. Desarrollo de un simulador de cadera incluyendo microseparación de prótesis totales de cadera. Tesis para obtener el grado de Maestro en Ciencias de la Ingeniería Mecánica con especialidad en Materiales. UANL, Junio de 2007.

[10] S. Affatato, W. Leardini, and M. Zavalloni. Hip Joint Simulators: State of the Art. 10th Ceramtec International Congress. 6th Symposia: Tribology (2006).

[11] B. Bermúdez-Reyes, F. J. Espinoza-Beltrán, I. Espitia-Cabrera and M .E. Contreras-García. Charaterization of HA/ZrO₂ – Base Bilayer on 316L Stainless Steel Substrates for Orthopedic Prosthesis Applications. Adv. in Tech. of Mat. and Mat. Proc. Vol. 9[2] 141-148 (2007).

[12] J.A. Ortega-Sáenz, M.A.L. Hernández-Rodríguez, A. Pérez-Unzueta, R. Mercado-Solis Development of a hip wear simulation rig including micro-separation. *Wear, Volume 263, Issues 7-12, 10 September 2007, Pages 1527-1532*

[13] Tertius Opperman. Tribological evaluation of joint fluid and development of a synthetic lubricant for use in hip simulators. Submitted in fulfillment of part of the requirements for of the Master's in Engineering, Building Environment and Information Technology. University of Pretoria. Pretoria. 09 November 2004

[14] O. Calonius, V. Saikko. Analysis of Relative Motion between Femoral Head and Acetabular Cup and Advances in Computation of the Wear Factor for the prosthesis hip Joint. Published in Acta Polytechnica. Vol. 43, 2003, No. 4, p. 43-54.

[15] Masaki Nakajima, Toshihiro Shimizu, Toshitaka Kanamori and Keiro Tokaji. Fatigue Crack Growth Beheviour of Metallic Biomaterials in a Physiologic Enviroment. Fatigue and Fracture of engineering Materials and Structures. 1998; 21: 35-45.

[16] J. Li, L. Hermansson, R. Söremark. High-strength biofuncional zirconia: mechanical properties and static fatigue behavior of zirconia-apatite composites. Journal of Material Science: Materials in Medicine 4(1993) 50-54.

[17] Qinglian Wang, Shirong Ge, Dekun Zhang. Nano-mechanical properties and biotribological behaviors of nanosized HA/partially-stabilized zirconia composites. Wear 259 (2005) 952-957.

[18] A. H. De Aza, J. Chevalier, G. Fantozzi, M. Schehl, R. Torrecillas. Crack growth resistance of alumina, zirconia and zirconia toughened alumina ceramics for joint prosthesis. Biomaterials 23(2002) 937-945.

[19] A. M. Patel y M. Spector. Tribological evaluation of oxidized zirconium using an articular cartilage counterface: a novel material for potential use in hemiarthoplasty. Biomaterials 18 (1997) 441-447.

[20] J. Chevalier, B. Calès,, J. M. Droulin, Y. Stefani. Ceramic-Ceramic bearing systems compared on different testing configuration. Bioceramics Vol 10. Proceedings of the 10th International Symposium on Ceramics in Medicine. Paris, France, October 1997.

[21] J. Caton, J. P. Bouraly, P. Reynaud and Z. Merabet. Phase transformation in zirconia heads after THA myth or reality?. 6th Ceramtec International Congress. 2002.

[22] H. G. Pfaff G. Willmann. Stability of Y-TZP zirconia. 2nd Ceramtec International Congress. 1997.

[23] C. Poconni, G. Maccauro, L. Pilloni, W. Buerger, F. Muratori, H. G. Richter. On the fracture of a zirconia balls head. Journal of Materials Science: Materials in Medicine 17(2006) 289-300.

[24] S. Affatato, M. Goldoni, M. testoni, A. Toni. Mixed oxides prosthetic ceramic ball head. Part 3: effect of the ZrO₂ fraction on wear on the ceramic on ceramic hip joint prostheses. A long-term in vitro wear study. Biomaterial 22 (2001) 717-723.

[25] G. Baroud and Willmann. Hydroxyapatite coating supports proximal load transfer a hip joint replacement. 3rd Ceramtec International Congress. 1998.

[26] Brian R. Lawn. Ceramic-based layer structures for biomechanical applications. Current Opinion in solid State and Materials Science 6 (2002) 229-235.

[27] G. Reinisch, K. P. Judmann and P. Pauschitz. Hip Simulator testing. 7th Ceramtec International Congress. 2002.

[28] G. Bergmann, F. Graichen and A. Rohlmann. Hip joint loadings during walking and running. Mearured in two patients. Journal of Biomechanics, Vol. 26, No. 3. 969-990 (1993). .

[29] Vesa Saikko, Olof Calonius. Slide track analysis of the relative motion between femoral head and acetabular cup in walking and in hip simulator. Journal of Biomechanics 35 (2002) 455-464.

[30] Olof Calonius. Tribology of Prosthetic Joints-Validation of Wear Simulator Methods. Dissertation for the degree of Doctor of Science in Technology. Helsinki University of Technology. 4th of October, 2002.

BIOINSPIRED CERAMIC MICROSTRUCTURES PREPARED BY FREEZING OF SUSPENSIONS

Qiang Fu[1]; Mohamed N. Rahaman[1]; B. Sonny Bal[2]; Fatih Dogan[1]
1. Department of Materials Science and Engineering, Missouri University of Science and Technology, Rolla, MO, USA.
2. Department of Orthopaedic Surgery, University of Missouri-Columbia, Columbia, MO, USA.

ABSTRACT

A generic method based on unidirectional freezing of suspensions was used for the preparation of porous three-dimensional bioceramics with microstructures mimicking those of natural materials such as nacre and bone. Microstructural manipulation in the freezing process was achieved by modifying the solvent composition of the suspension. The fabricated constructs had high toughness, high strain for failure, and strain rate sensitivity, typical of many natural materials. The porous constructs could be applied as scaffolds for bone repair and regeneration.

INTRODUCTION

Natural materials are renowned for their unique combination of outstanding mechanical properties and exquisite microstructure. Cortical (or compact) bone, teeth, and shells are biological inorganic–organic composites, composed of minerals such as calcium phosphate or calcium carbonate, with high mechanical strength and toughness[1-6]. Trabecular (or cancellous) bone, cork and wood are biological cellular materials with high specific stiffness (stiffness per unit weight) and specific strength[7-10]. The outstanding mechanical properties of biological mineral composites and cellular materials are attributed to the complex hierarchical architecture at several levels[7-9,11]. With inspiration from the biological materials, scientists are developing new materials and structures using both conventional and molecular-based approaches[6,12,13]. The conventional biomimetic approach utilizes synthetic materials and conventional processing methods to produce structures that mimic those of biological materials, such as laminated ceramic composites inspired by the structure of shells[3-5]. Molecular-based approaches utilize self-assembly of the components at the molecular or nanoscale level, and the development of hierarchical structures[6,14], such as a self-assembled mesoscale networks and ordered quantum dots[15,16]. Organic molecules, which self-assemble into a pre-organized template in aqueous environment, have been reported to play an important role in the manipulation of nucleation, growth, microstructure, and the properties of the natural materials[17-19].

Freezing of aqueous suspension is a traditional biomimetic approach which utilizes ice as a template for fabricating porous inorganic and polymeric structures[20-22]. Ceramic composites fabricated by directionally freezing can mimic the structure of nacre[21]. The lamellar structure with aligned pore morphology results from the unidirectional growth of ice crystals which expels the fine particles from the growing crystals. So far, the structures developed by freezing aqueous suspensions have been confined to the production of lamellar –type microstructures because of the hexagonal crystal structure of ice, which determines the pore morphology of the fabricated materials.

Here we show how the structure of ceramics can be manipulated into diverse and organized long-range architectures by using different organic solvents to modify the freezing behavior of the aqueous solvent in a suspension. The comparison of the synthetic ceramics and natural materials was also investigated.

MATERIALS AND METHODS

Materials

Hydroxyapatite (HA) powder (Alfa Aesar, Haverhill, MA), with an average particle size of <0.5 μm determined by a laser diffraction particle size analyzer (Model LS 13 320; Beckman Coulter Inc., Fullerton, CA), was used in the present work. The aqueous suspensions contained 5–20 weight percent (wt%) HA particles, 0.75 wt% of a dispersant (Dynol 604; Air Products & Chemicals Inc., Allentown PA), and 1.5 weight percent (wt%) of an organic binder, poly(vinyl alcohol), PVA, (DuPont Elvanol® 90-50, DuPont, DE), based on the dry weight of the HA powder. The suspensions were prepared by ball-milling the mixture for 48 h in polypropylene containers using Al_2O_3 grinding media. Aqueous suspensions containing 5–20 wt% of glycerol (Fisher Scientific, Pittsburg, PA) or 40–70 wt% dioxane (Fisher Scientific) were prepared using the same procedure.

Freeze casting of suspensions

Freeze casting was performed by pouring the suspensions into polyvinyl chloride (PVC) tubes (~10 mm internal diameter ~ 20 mm long) placed on a cold steel substrate kept at –20°C in a freeze dryer (Genesis 25 SQ Freeze Dryer, VirTis Co., Gardiner, NY). To promote unidirectional freezing, the PVC tubes were insulated with polyurethane foam to reduce heat transfer from the surrounding environment. The frozen constructs were subjected to a vacuum of 4 Pa for 48 h in a freeze dryer (Genesis 25 SQ Freeze Dryer) to cause sublimation of the frozen solvent. The freeze-dried constructs were sintered in air, for 3 h at 1375°C (heating and cooling rate = 3°C/min) to densify the solid phase of the constructs.

Materials characterization

The viscosity of the suspensions as a function of shear rate was measured using a rotating cylinder viscometer (Model VT500; Haake Inc., Paramus, NJ). Scanning electron microscopy, SEM (Hitachi S-4700, Hitachi Co., Tokyo, Japan) was used to observe the microstructures of the fabricated constrcuts. The compressive strength of cylindrical samples (8 mm in diameter ~ 16 mm) in the directions parallel and perpendicular to freezing was measured according to ASTM-C773 using an Instron testing machine (Model 4204, Norwood, MA) at a crosshead speed of 0.5 mm/min. Eight samples were tested, and the average strength and standard deviation were determined. To investigate the effect of loading (strain) rate on the compressive mechanical response, testing was also performed at cross-head speeds of 0.05 and 5 mm/min.

RESULTS AND DISCUSSION

Unidirectional freezing of suspensions provides a generic approach for mimicking the complex structure of some natural materials. By controlling the solvent composition of the suspension and the freezing parameters, constructs with microstructures mimicking those of nacre, cancellous bone and cork were prepared (Fig. 1). Freezing of aqueous hydroxyapatite suspension leads to a lamellar-type microstructure (Fig. 1a), similar to the structure of nacre (Fig. 1b). By modifying the freezing behavior of the aqueous solvent of the suspensions, through additions of an organic solvent such as glycerol or dioxane, constructs that mimic the structure of human cancellous bone (Fig. 1c,d), and cuttlefish bone (Fig. 1e,f) are obtained. The microstructures of the freeze-cast constructs can be classified into three general types: lamellar, cellular, and rectangular, with pore widths of several micrometers to >100 μm. This diversity in the microstructure is attributed to the modification of the H–bonding, and, hence, the freezing behavior of the aqueous solvent, by the addition of organic solvents such as glycerol or dioxane.

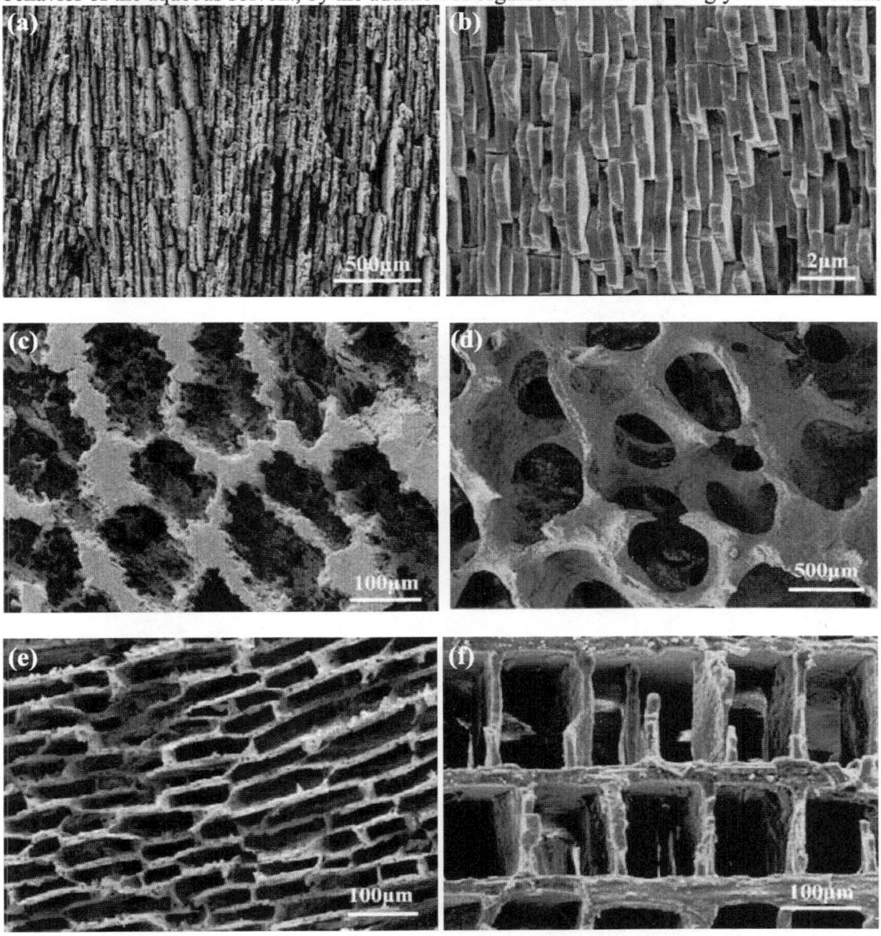

Figure 1 SEM images showing a microstructural comparison of hydroxyapatite constructs prepared in the present work (left) with natural materials (right). (a) Constructs prepared from aqueous suspension; (b) nacre; (c) constructs prepared from aqueous suspension with 23 mol% (60 wt%) dioxane; (d), cancellous bone; (e) constructs prepared from aqueous suspension with 1.0 mol% (5 wt%) glycerol; (f) cuttlefish bone taken from Ref. 29.

In addition to mimicking the microstructures of natural materials, the constructs prepared in the present work show the mechanical response characteristic of tough natural materials. When loaded in compression along the direction parallel to the freezing direction, hydroxyapatite constructs prepared from aqueous suspensions (20 vol% particles) show a bell-shaped response (Fig. 2a), not significantly unlike the response of nacre when loaded in compression parallel to the layers (Fig. 2b)[30]. In comparison, hydroxyapatite constructs prepared from aqueous suspensions (10 vol% particles) containing dioxane (23 mol%) show a mechanical response (Fig. 3a) similar to that of cancellous bone (Fig. 3b) and wood [8,31]. The stress vs. strain response for the hydroxyapatite constructs (Fig. 3(a) can be divided into three distinct regimes: a linear region, a stress plateau region, and a densification regiion. The linear region corresponds to the bending of the pore walls, whereas the stress plateau region results from the crushing of the pore walls. Densification in the third region results from collapse of the structure.

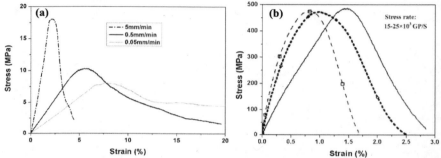

Figure 2 Stress vs. strain response in compression: (a) hydroxyapatite constructs prepared in the present work from aqueous suspension (20 vol% particles) at different strain rates in the direction parallel to the freezing direction; (b) results for nacre, taken from Ref. 30.

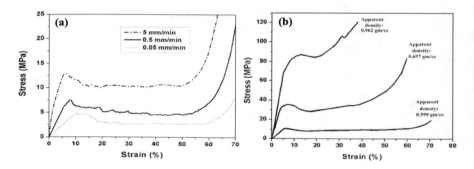

Figure 3 Stress vs. strain response in compression: (a) hydroxyapatite constructs prepared from aqueous suspension (10 vol% particles) with X mol% (60 wt%) dioxane, in the direction parallel to the freezing direction; (b) results for cancellous bone taken from Ref. 31.

The mechanical strength of the cellular hydroxyapatite constructs, prepared from aqueous suspensions with dioxane (Fig. 3a), can be well described by the Gibson and Ashby model for cellular materials8. According to the model, the compressive crushing strength of brittle foams is given by:

$$\frac{\sigma_{cr}}{\sigma_{fs}} = C\left(\frac{\rho}{\rho_o}\right)^{3/2} \frac{1+(t_i/t)^2}{\sqrt{1-(t_i/t)^2}} = C(1-P)^{3/2} \frac{1+(t_i/t)^2}{\sqrt{1-(t_i/t)^2}} \tag{1}$$

where σ_{cr} is the compressive crushing strength of the foam, σ_{fs} is the modulus of rupture of the solid struts, C is a constant of proportionality, ρ and ρ_o are the densities of the porous cellular material and the fully dense solid phase, respectively, P is the porosity of the cellular material, and t_i/t is the ratio of width of the box-like void in the struts to the thickness of the solid strut. In the case of brittle foams[8], C = 0.2. For hydroxyapatite, the reported values for σ_{fs} are in the range 115–200 MPa[32]. In the present work, the struts were almost fully dense, so $t_i/t = 0$, and P $\approx$ 0.65. Using these values, the strength predicted by the model is in the range 5–8 MPa, which is in good agreement with the measured value of 7.5 MPa. The crushing strength of hydroxyapatite constructs was determined as peak value in the strain-stress curve (Fig. 5a) when compressed under a strain rate of 0.5 mm/min along the pore direction.

Although the solid phase of the constructs prepared in this work consist of a single-phase inorganic (ceramic) material, the stress vs. strain response does not show the brittle behavior characteristic of ceramics, with just an elastic response and failure at low strain (typically <0.1%). The mechanical response of the constructs show high strain tolerance (several percent strain), with a high strain to failure (Figs. 2 and 3). The toughness of the constructs is attributed to crack deflection, branching and blunting mechanisms, similar to those described for nacre[30,33]. Similar to the behavior of bone and other natural materials[34,35], the mechanical response of the constructs prepared in this work is sensitive to the rate of loading (Figs. 2 and 3). The strength increases with higher loading rate, but the strain to failure decreases. Since the solid phase (hydroxyapatite) is brittle, the strain rate sensitivity results from the microstructures developed by the freezing route.

CONCLUSIONS

The constructs prepared in this work could provide novel substrates for in vitro cell culture experiments, as well as novel scaffolds for bone repair and regeneration. The method is applicable not just to hydroxyapatite described in the present work, but also to other biocompatible and bioactive inorganic materials, such as bioactive glass. It is well known that cells can respond to the shape of their environment. The diverse structures obtained with the freezing approach can provide new substrates for evaluating the influence of pore architecture on in vitro cell proliferation.

REFERENCES
[1]S. Weiner, H.D.Wagner, The material bone: structure-mechanical function relations. Annu.Rev. Mater. Sci. 28, 271-98 (1998).
[2]H. Peterlik, P. Roschger, K.Klaushofer, P.Fratzl, From brittle to ductile facture of bone. Nature Mater. 5, 52-5 (2006).
[3]I.A.Aksasy, M.Trau, S.Manne, I.Honma, N. Yao, L.Zhou, P.Fenter, P.M. Eisenberger, S.M. Gruner, Biomimetic pathways for assembling inorganic thin films. Science 273, 892–98 (1996).

[4]A. Sellinger, et al. Continuous self-assembly of organic–inorganic nanocomposite coatings that mimic nacre. Nature 394, 256–60 (1998).

[5]Tang, Z., Kotov, N.A., Magonov, S., Ozturk, B. Nanostructured artificial nacre. Nature Mater. 2, 413-18 (2003)

[6]M.A. Meyers, P. Chen, A.Y. Lin, Y.Seki, Biological materials: structure and mechanical properties. Prog. Mater. Sci. 53, 1-206 (2008)

[7]M.F.Ashby, L.J.Gibson U. Wegst, R.Olive, The mechanical properties of natural materials. II. Microstructure for mechanical efficiency. Proc. R. Soc. Lond. A 450, 141-62 (1995)

[8]L.J.Gibson, M.F.Ashby, Cellular solids: structure and properties. 2nd ed. Cambridge: Cambridge University Press (1997)

[9]P.Fratzl, R.Weinkamer, Nature's hierarchical materials. Prog. Mater. Sci. 52, 1263-334 (2007)

[10]J.Keckes, I.Burgert, K.Frühmann, M.Müller, K.Kölln, M.Hamilton, M.Burghammer, S.Roth, S.Stanzl-Tschegg, P.Fratzl, Cell-wall recovery after irreversible deformation of wood. Nature Mater. 2, 810-4 (2003)

[11]J.D. Curry, The design of mineralized hard tissues for their mechanical functions. J Exp. Biol. 202, 3285-94 (1999).

[12]G. Meyer, Rigid biological systems as models for synthetic composites. Science 310, 1144-7 (2005)

[13]C. Sanchez, H.Arribart, M.M.G. Guille, Biomimetism and bioinspiration as tools for the design of innovative materials and systems. Nature Mater. 4, 277-88 (2005)

[14]S. Mann, Molecular tectonics in biomineralization and biomimetic materials chemistry. Nature 365, 499-505 (1993)

[15]N.Bowden, A.Terfor, J.Carbeck, G.M.Whitesides, Self-assembly of mesoscale objects into ordered two-dimensional arrays. Science 276, 233-5 (1997)

[16]S.W.Lee, C.Mao, C.E.,lynn, A.M. Belcher, Ordering of quantum dots using genetically engineered viruses. Science 296, 892-5 (2002)

[17]S. Mann, Molecular tectonics in biomineralization and biomimetic materials chemistry. Nature 365, 499-505 (1993).

[18]S. Mann, G.A.Ozin, Synthesis of inorganic materials with complex form. Nature 382, 313-8 (1996).

[19]S.I.Stupp, P.V.Braun, Molecular manipulation of microstructure: biomaterials, ceramics, and semiconductors. Science 277, 1242-8 (1997).

[20] H.Zhang, I.Hussain, M.Brust, M.F.Butler, S.P.Rannard, A.I.Cooper, Aligned two- and three-dimensional structures by directional freezing of polymers and nanoparticles. Nature Mater. 4, 787-93 (2005)

[21]S.Deville, E.Saiz, R.K.Nalla, A.P.Tomsia, Freezing as a path to build complex composites. Science 311, 515-8 (2006)

[22]Q. Fu, M.N. Rahaman, F.Dogan, B.S.Bal, Freeze casting of porous hydroxyapatite scaffolds - I. Processing and general microstructure. J. Biomed. Mater. Res. B 86B 125-35 (2008)..

[23]J.L. Dashnau, N.V.Nucci, K.A.Sharp, J.M.Vanderkooi, Hydrogen bonding and cryoprotective properties of glycerol/water mixtures. J. Phys. Chem. B 110, 13670-13677 (2006)

[24]S.W.Sofie, F.Dogan, Freeze casting of aqueous alumina slurries with glycerol. J. Am. Ceram. Soc. 84, 1459-64 (2001)

[25]Y.G.Wu, M.Tabata, T.Takamuku, A local solvent structure study on 1,4-dioxane-water binary mixtures by total isotropic Rayleigh light scattering method. J. Mol. Liq. 94, 273-82 (2001)

[26]J.Mazurkiewicz, P.Tomsasik, Why 1,4-dioxane is a water-structure breaker. J. Mol. Liq. 126, 111-6 (2006)

[27]D.R.Uhlmann, B. Chalmers, K.A.Jackson, Interaction between particles and a moving ice-liquid interface. J. Appl. Phys. 35, 2986-93 (1964)

[28]C.Kőrber, G.Rau, M.D.Cosman, E.G. Cravalho, Interaction of particles and a moving ice-liquid interface. J. Cryst. Growth 72, 649-62 (1985).

[29]J.H.G.Rocha, A.F.Lemos, S.Agathopoulos, P.Valério, S.Kannan, F.N.Oktar, J.M.F.Ferreira, Scaffolds for bone restoration from cuttlefish. Bone 37, 850-7 (2005)

[30]. R.Menig, M.H.Meyer, M.A.Meyers, K.S.Vecchio, Quasi-static and dynamic mechanical response of halliotis rufescens (abalone) shells. Acta Mater. 48, 2383-98 (2000).

[31]W.C. Hayes, D.R.Carter, Postyield behavior of subchondral trabecular bone. J. Biomed. Mater. Res. 10, 537-44 (1976).

[32]. Hench, L. L. Bioceramics. J. Am. Ceram. Soc., 81,1705-28 (1998).

[33]Q.Fu, M.N. Rahaman, F.Dogan, B.S. Bal, Freeze casting of porous hydroxyapatite scaffolds - II. Sintering, microstructure and mechanical behavior. J. Biomed. Mater. Res. B 86B 514-22 (2008)

[34]J.H. McElhaney, Dynamic response of bone and muscle tissue. J. Appl. Physiol. 21, 1231-6 (1966).

[35]D.R.Carter, Bone compressive strength: the influence of density and strain rate. Science. 194,1174-6 (1976).

MECHANICAL PROPERTIES MODELING OF POROUS CALCIUM PHOSPHATES CERAMICS

François Pecqueux ; Franck Tancret ; Nathalie Payraudeau
Université de Nantes - Laboratoire Génie des Matériaux et Procédés Associés – Polytech' Nantes
BP 50609 – 44306 Nantes Cedex 3 – France

Jean-Michel Bouler
INSERM UMR 791 - Ingénierie Ostéo-Articulaire et Dentaire – Université de Nantes
1, place Alexis Ricordeau - BP84215 – 44042 Nantes Cedex 1 – France

ABSTRACT

Macroporous Biphasic Calcium Phosphate bioceramics, for use as bone substitutes, have been fabricated by soft chemical hydrolysis, cold isostatic pressing and conventional sintering, using naphthalene particles as a porogen to produce macropores. The resulting ceramics are biphasic materials made of hydroxyapatite and β-tricalcium phosphate containing various macroporosities and microporosities. Compression tests and three-point bending toughness tests were performed on samples covering the widest attainable ranges of porosities (53 different macroporosity/microporosity couples). Results are compared to previously proposed analytical models and hypotheses. The models describe the evolution of mechanical properties of a porous material, taking into account the separate influences of both macroporosity and microporosity. The strength model also considers macropores as critical flaws. The main results are the following: (i) the models are modified to improve the fitting with experiment, (ii) the macropores are actually identified as being linked to the critical flaws and (iii) the previous hypothesis is rejected, as the critical flaw size appears to increase with macroporosity. This phenomenon is interpreted by the existence of clusters of macropores, acting as critical flaws, which size increases with macroporosity.

INTRODUCTION AND BACKGROUND

Biphasic Calcium Phosphates (BCP) are commonly used as bone substitutes. They are porous mixtures of Hydroxyapatite, or HA, of formula ($Ca_{10}(PO_4)_6(OH)_2$) and β-Tri-Calcium Phosphate, or β-TCP, of formula ($Ca_3(PO_4)_2$). These synthetic bone grafts require a specific porous structure combining macroporosity and microporosity (Figure 1) to be degraded by bone cells and therefore replaced by natural bone[1]. The macroporosity (p_{macro}) is the ratio of the macropores volume over the specimen one, while the microporosity (p_{micro}) is the volume of the pores reminding between the ceramic matrix grains after sintering.
To improve the mechanical properties of these materials, it is foremost necessary to be able to describe their variation as a function of both porosity types. Several studies were led on the mechanical behavior of brittle alveolar materials. Among others, in 1991, Wagh et al.[2] proposed an interesting model description of the Young's modulus as a function of porosity that holds in the case of closed isolated pores (corresponding, in the present case, to macropores):

$$E = E_0.(1 - p_{macro})^m \quad (1)$$

with p_{macro} the specimen porosity (in our case the macroporosity), m a parameter depending of the porosity morphology, and E_0 the Young's Modulus of the fully dense material, *i.e.* for $p_{macro} = 0$.

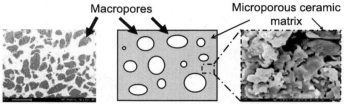

Figure 1. Scanning electron micrographs and schema of the BCP porous structure.

In 1993, the same group of authors asserted that this kind of model construction could be used also for toughness and strength[3] provided that both the fracture strain and the critical flaw size were independent of porosity. This approach was corrected later by Arató[4], who showed that under such conditions the same function of porosity could be used for the relative Young's modulus, relative toughness and relative fracture strength.

Other models have been proposed to describe the variation of mechanical properties as functions of the porosity resulting from the incomplete densification by sintering of a pressed powder (corresponding, in the present case, to the microporous matrix of the ceramic). Particularly, Jernot *et al.* published in 1982, a geometrical based model for this type of porosity[5]:

$$E = E_0.[N_C.(1 - p_{micro}) - (N_C - 1).(1 - p_{micro})^{2/3}] \tag{2}$$

with p_{micro} the residual porosity (in our case the microporosity) and N_C the average number of closest neighbors for each grain in the pressed powder before sintering.

In 2006, Tancret *et al.* made the hypothesis that the respective effects of macropores and micropores could be decorrelated as a consequence of their significant difference in size (almost two orders of magnitude). Thus, they combined these two models (Equations 1 and 2) into a single one to describe the Young's modulus of BCP ceramics containing both macropores and micropores[6]:

$$E = E_0.[N_C.(1 - p_{micro}) - (N_C - 1).(1 - p_{micro})^{2/3}].(1 - p_{macro})^m \tag{3}$$

The rules of Wagh *et al.*[3] and Arató[4] were applied, which led to the definition of a toughness model:

$$K_{IC} = K_{IC0}.[N_C.(1 - p_{micro}) - (N_C - 1).(1 - p_{micro})^{2/3}].(1 - p_{macro})^m \tag{4}$$

with K_{IC0} the toughness of the fully dense material, *i.e.* with $p_{micro} = p_{macro} = 0$.

Based on the observation that the calculated critical flaw size was nearly independent of porosity, and of the order of the size of macropores, it was assumed that one of them was the critical flaw[6]. Given the large number of calibrated macropores in each specimen, it could be considered that the critical flaw was always of similar size and shape. Writing the classical relation between fracture toughness, strength and critical flaw size and shape:

$$\sigma_r = \frac{K_{IC}}{Y.\sqrt{a_C}} \text{ with Y and } a_C \text{ constant} => \sigma_{r0} = \frac{K_{IC0}}{Y.\sqrt{a_C}} \tag{5}$$

with Y a parameter associated to the geometry of the critical flaw, a_C its radius and σ_{r0} the strength of the fully dense material, *i.e.* with $p_{micro} = p_{macro} = 0$. Then, this led to the definition of a strength model:

$$K_{IC} = K_{IC0}.f(p_{macro}; p_{micro}) => \sigma_r = \frac{K_{IC}}{Y.\sqrt{a_C}} = \frac{K_{IC0}.f(p_{macro}; p_{micro})}{Y.\sqrt{a_C}} = \sigma_{r0}.f(p_{macro}; p_{micro}) \tag{6}$$

$$\sigma_r = \sigma_{r0}.[N_C.(1-p_{micro}) - (N_C-1).(1-p_{micro})^{2/3}].(1-p_{macro})^m \tag{7}$$

As all the specimens are fabricated from the same powders, the values of the parameters m and N_C should be the same in the toughness and the strength models.

The aims of this work were to validate this model by testing the mechanical properties over the widest attainable range of macroporosities and microporosities and to confirm the above-mentioned hypotheses, *i.e.*:

- The critical flaw in such porous materials is always a macropore.
- The analytical models (Equations 4 and 7) can describe the toughness and the compressive strength with the same parameters values, whatever are the amounts of both macroporosity and microporosity.

MATERIALS AND TESTING

Fabrication of materials

The Biphasic Calcium Phosphate (BCP) used to fabricate all the mechanical testing specimens was synthesized in our laboratory to make sure that its chemical composition and its granulometry are under control. Two batches of CDA powder (thereafter named CDA-A and CDA-B) were synthesized by hydrolyses of two commercial powders of Di-Calcium Phosphate Di-hydrated (DCPD) in an aqueous solution of NH_4OH heated at 70°C and maintained under vigorous stirring during 6 h. The starting pH and the stoichiometry of the reactants were chosen to insure a final atomic Ca/P ratio around 1.52. The solution was then filtered and dried at 120°C for one night. Sintering of the CDA powders leads to the formation of mixtures of HA and βTCP, the proportions of which were determined by classical X-ray Diffraction. The sintering of CDA-A leads to a BCP-A composed of 15% HA and 85% β-TCP while CDA-B leads to a BCP-B composed of 20% HA and 80% β-TCP. Infrared Spectroscopy was used to check the purity of the powders.

The CDA powders were mixed in a Turbula Shaker-mixer with various proportions of naphthalene particles (sifted between 200 and 600 μm) to fabricate plates by cold isostatic pressing under 140 MPa. The plates are placed in a kiln at 80°C during 48 h to sublimate the naphthalene particles and obtain pressed CDA blocks containing various macroporosities (from 0% to 53%). The material was pressureless sintered in air for 8 h at temperatures from 850°C to 1250°C. These sintering cycles reached a temperature sufficient to transform 100% of CDA into BCP but with a matrix (the ceramic material lying between the macropores) containing a variable microporosity from 2% to 50%.

To evaluate their macroporosity by ceramography[7], small blocks where cut in each plate, impregnated with an epoxy resin under vacuum and polished with SiC papers down to P4000 and then with diamond suspension down to 1μm in grain size. Macroporosity is then measured by quantitative image analysis on polished cross-sections. Total porosity is estimated through the apparent density of the samples; the microporous volume is then obtained by subtracting the macroporous volume and the solid volume from the total volume.

Mechanical properties testing

Compression tests were performed with a universal mechanical testing machine on parallelepipedic blocks of about 7 mm×7 mm×11 mm cut out from each plate.
Fracture toughness was tested in three-point bending on chevron-notched parallelepipedic bars of about 8 mm ×10 mm×50 mm cut from each plate of batch B following a procedure described by Dlouhy et al.[8] (Figure 2).

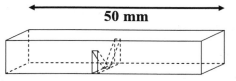

50 mm

Figure 2. Schema and fractography of a chevron notched bar.

RESULTS AND DISCUSSION

Fitting procedure

Several specimens of each macroporosity/microporosity couple have been mechanically tested. The means of these measurements have been plotted in different diagrams (Figures 3, 5, 7 and 9), but the error bars have not been displayed for clarity. The analytical models parameters are then adjusted to obtain the best fit of the experimental data, by minimizing the sum of the overall relative difference between the measured values and the calculated ones.
At the beginning of the fitting procedure, we found that it was necessary to modify the previous formulas (Equations 3, 4 and 7) by adding a constant term (X_{min}) in each model (see Equations 8 and 9 below). This minimum value corresponds to the mechanical property that would be measured on specimens cut in a plate at the start of sintering, i.e. after the CDA => HA + β-TCP phase transformation and after a possible grain rearrangement, but before shrinkage begins.

Toughness analytical model

$$K_{IC} = K_{IC0}.[N_C.(1 - p_{micro}) - (N_C - 1).(1 - p_{micro})^{2/3}].(1 - p_{macro})^m + K_{IC\,min} \qquad (8)$$

with $K_{IC0}+K_{ICmin}$ the toughness of the fully dense material, i.e. with $p_{micro} = p_{macro} = 0$.
Numerous toughness tests have been performed on chevron notched bars cut out from 14 different plates of the material fabricated with the CDA-B powder (one or two bars from each plate). The best description of the collected data seems to be for K_{IC0} = 1.53 MPa√m, K_{ICmin} = 0.05 MPa√m, N_C = 4.56 and m = 2.47 (Figure 3). These values are close to classical

values reported for ceramic materials (K_{IC} less than several MPa√m for dense materials, N_C between 5 and 7 and m close to 2)[2,4].

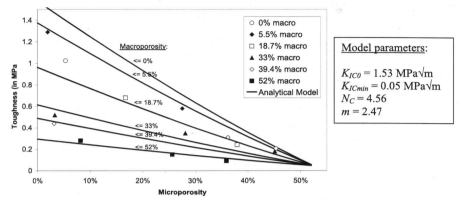

Figure 3. Experimental and modeled toughness as function of porosities (batch B).

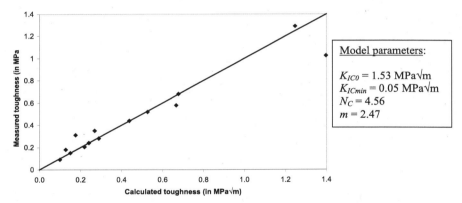

Figure 4. Comparison between experimental and calculated values for toughness (batch B).
Another way to present results is to report measured values as a function of calculated ones. A rather good agreement is found between the measured and the calculated values (as shown on Figure 4).

Compression strength analytical model

$$\sigma_r = \sigma_{r0}.[N_C.(1-p_{micro})-(N_C-1).(1-p_{micro})^{2/3}].(1-p_{macro})^m + \sigma_{r\min} \quad (9)$$

with $\sigma_{r0}+\sigma_{rmin}$ the intrinsic compressive strength of the fully dense material, *i.e.* with $p_{micro} = p_{macro} = 0$.

Materials made with the CDA-B powder, containing different couples of porosities have been tested to measure their compressive strength. The blocks used for the compression tests are much smaller than the bars used for toughness tests, as a consequence it was possible to cut and test more little blocks for each plate (between 3 and 7 from each plate). The best description of the collected data seems to be for $\sigma_{r0} = 225$ MPa, $\sigma_{rmin} = 1$ MPa, $N_C = 4.56$ and $m = 4.58$ (Figure 5).

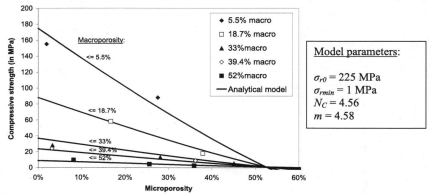

Figure 5. Experimental and modeled compressive strength as function of porosities (batch B).

A very good agreement is found between the measured and the calculated values (as shown on Figure 6).

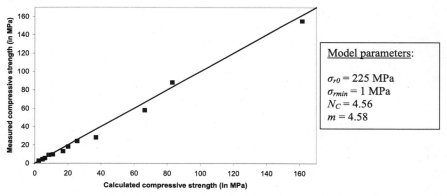

Figure 6. Comparison between experimental and calculated values for compressive strength (batch B).

Numerous compression tests have also been performed on blocks cut in 39 different plates of the material fabricated from the CDA-A powder. The best description of the collected data seems to be for $\sigma_{r0} = 225$ MPa, $\sigma_{rmin} = 0.8$ MPa, $N_C = 4.56$ and $m = 4.58$ (Figure 7).

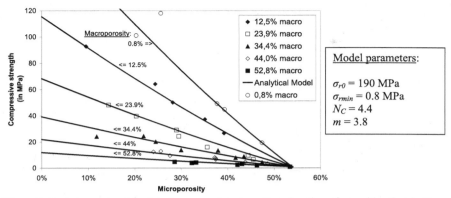

Figure 7. Experimental and modeled compressive strength as function of porosities (batch A).

An excellent agreement is found between the measured and the calculated values (as shown on Figure 8) as soon as the materials contain macropores. Nevertheless, the model is not able to fit the data with the same values of N_C and m when the ceramics contain no macropore, as shown on Figures 8 and 9.

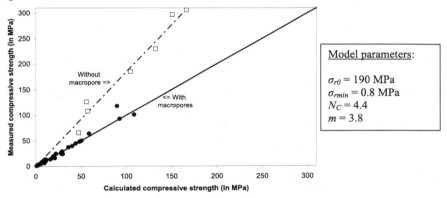

Figure 8. Comparison between experimental and calculated values for compressive strength (batch A).

Indeed, the compressive strength the 0% macroporosity materials seems about twice higher than the calculated one. However it is possible to obtain a very good regression of the experimental compressive strength of the material containing no macropore by changing the parameter value σ_{r0} to 465 MPa and using $p_{macro} = 0$. This means that the analytical model is still able to describe the behavior of the material but that the critical flaw is smaller in the materials containing no macropore than in the macroporous ceramics. This observation tends to confirm that the critical flaw is a macropore as soon as macropores are present in the material.

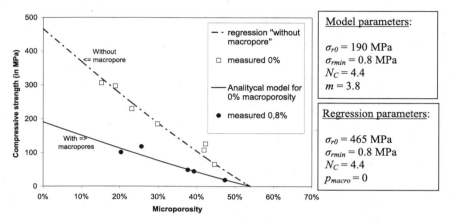

Figure 9. Experimental and modeled compressive strength as function of porosities (batch A).

Other experiments were performed to try to comfort this hypothesis; little bars containing no macropore or a single naphthalene crystal in different locations were pressed (see Figure 10). After heat treatment to create single macropores and sintering, the bars were broken on a three-point bending setup. For every bar containing no macropore or a macropore situated in a low stress area, the fracture occurred at the vertical of the central loading point, whereas for the bars containing a single macropore in a highly stressed area (*i.e.* close to the vertical of the loading point and close to the lower surface of the specimen), the fracture of the specimen occurred through the macropore even if not exactly below the loading point. The shift of the fracture location in this case seems to indicate that in these specimens, the rupture was due to an initial crack starting from the macropore. This observation tends also to confirm that a single macropore is always the critical flaw in these specimens.

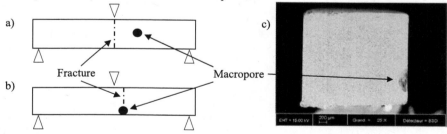

Figure 10. a) and b) Setup of 3-point bending tests, c) Fractography of a specimen

Besides, in the case of the materials fabricated from the CDA-B powder, it can be noted that the best fits displayed in Figures 3 and 5 have been obtained with the same value of N_C for toughness and strength, but with different values of m (2.47 and 4.58, respectively). N_C values are equal, confirming that the influence of microporosity is the same for the two mechanical properties, whereas the apparent m value is much higher for the strength model than for the toughness one. However, the definition of the models indicates that all mechanical properties should be described with a single set of parameters N_C and m. In particular, the same parameters should be used in the strength and in the toughness models if the critical flaw size and shape are independent of porosity, as stated in the introduction and expressed by Equations 4 to 7. In our case, the initial hypothesis was that a macropore, of "constant" size (typically hundreds of μm) and shape is always the critical flaw[6].

Nevertheless, if we calculate for our materials the quantity $(K_{IC}/\sigma_r)^2$, which is proportional to the critical flaw size (as defined in Equation 5), it appears that this size globally increases with macroporosity, as shown on Figure 11.

P_{macro}	P_{micro}	σ_{meas} MPa	K_{ICmeas} MPa√m	$(K_{IC}/\sigma_r)^2$ µm
0.0%	45.3%	29	0.205	50
3.1%	27.5%	88	0.58	43
7.9%	1.9%	155	1.29	69
16.6%	16.7%	58	0.68	137
20.8%	37.9%	18	0.24	178
31.0%	45.1%	5.5	0.18	1071
34.4%	28.1%	13	0.35	725
35.6%	3.3%	28	0.52	345
38.2%	36.2%	9	0.31	1186
40.7%	3.1%	24	0.44	336
48.3%	8.1%	9.5	0.28	869
53.6%	25.6%	4.3	0.15	1217
54.5%	35.9%	2.7	0.09	1111

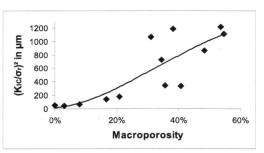

Figure 11. Variation of $(K_{IC}/\sigma_r)^2$ as a function of macroporosity.

This observation negates the above-mentioned hypothesis, *i.e.* the critical flaw is not constant in size. From a mathematical point of view, Equation 5 rewrites:

$$\sigma_r = \frac{K_{IC0} \cdot f(p_{macro}; p_{micro})}{Y(p_{macro}) \cdot \sqrt{a_C(p_{macro})}} = \sigma_{r0} \cdot g(p_{macro}; p_{micro}) \tag{10}$$

This is consistent with a higher "apparent" value of m for strength than for toughness. Indeed, if a_C increases with p_{macro}, the strength will decrease faster with macroporosity than toughness (function g decreases faster with p_{macro} than function f), which could also be fitted using higher values of m.

A simple explanation can be given to this increase in the size of the critical flaw with macroporosity: When there is only a limited number of macropores in the specimen (low macroporosity), macropores are rather scattered in the ceramic matrix and do not interact with each other (Figure 12a). When macroporosity increases, the average distance between two macropores decreases. Additionally, the probability to find groups of macropores in some areas increases (Figure 12b). Such groups of macropores can constitute weakened zones of the material, and act as a new enlarged critical flaw instead of a single macropore (Figure 12c). As macroporosity increases, these groups of macropores will statistically become larger, which corresponds to the observed progressive increase in critical flaw size.

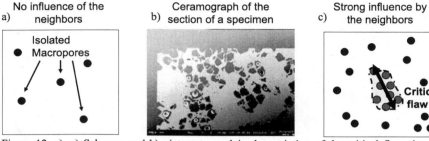

Figure 12. a), c) Schemas and b) picture to explain the variation of the critical flaw size with macroporosity.

CONCLUSION AND PERSPECTIVES

An important study has been undertaken on the variation with porosity of the mechanical properties of BCP ceramics exhibiting real implant structures. The variations of toughness and compressive strength of the material have been measured and compared with analytical models. The latter have also been adapted to the presently investigated ceramics. The results confirm that the behavior of the BCP ceramics can be described by these analytical models as functions of macroporosity and microporosity separately. However, the initial hypothesis used to build the strength model, *i.e.* that the critical flaw is always an isolated macropore, has been rejected. A new hypothesis is proposed, suggesting that a group of macropores can act as the critical flaw, which is consistent with experimental observations and measurements, although no direct evidence exists so far.

As a consequence, one of the perspectives of this study is to introduce voluntarily clusters of macropores of different sizes and measure their influence on the mechanical properties.

Another test campaign will also be performed to measure the variation of the Young's modulus of the BCP ceramic as a function of the two separate porosities and try to validate the construction of the initial model proposed by Tancret *et al.*[6]

These validated analytical models should be useful in view of forecasting the mechanical properties of a macroporous and microporous brittle material on the basis of only a few tests. In the end, a better understanding of the influence of microstructure on mechanical properties will enable the design of bioceramics with improved performance.

REFERENCES

[1]J.-M. Bouler, M. Trecant, J. Delecrin, J. Royer, N. Passuti, G. Daculsi, Macroporous biphasic calcium phosphate ceramics: influence of macropore diameter and macroporosity percentage on bone ingrowth, J. Biomed Mater. Res., **32**, 603-09 (1996)

[2]A.S. Wagh, R.B. Poeppel, J.P. Singh, Open pore description of mechanical properties of ceramics, J. Mater. Sci., **26**, 3862-68 (1991)

[3]A.S. Wagh, J.P. Singh, Dependence of ceramics fracture properties on porosity, R.B. Poeppel, J. Mater. Sci., **28,** 3589-93 (1993)

[4]P. Arató, Comment on « Dependence of ceramics fracture properties on porosity », J. Mater. Sci. Letters, **15,** 32-33 (1996)

[5]J.P. Jernot, M. Coster, J.L. Chermant, Model to describe the elastic modulus of sintered materials, Phys. Stat. Sol. (a), **72,** 325-32 (1982)

[6]F. Tancret, J.-M. Bouler, J. Chamousset, L.-M. Minois, Modelling the mechanical properties of microporous and macroporous biphasic calcium phosphate bioceramics, J. Eur. Ceram. Soc., **26,** 3647-56 (2006)

[7]J.-J. Friel, Practical Guide to Image Analysis., ASM International (1992)

[8]I. Dlouhy, M. Holzmann, J. Man, L. Valka, The use of chevron notched specimen for fracture toughness determination, Metal. Mater, **32,** 3-13 (1994)

BONE CEMENT REINFORCED WITH ZIRCONIUM OXIDE PARTICLES

H. H. Rodríguez and M. C. Piña*

Universidad Nacional Autónoma de México, Instituto de Investigaciones en Materiales
Ciudad Universitaria, Circuito Exterior s/n, México 04510 D.F. México
*mcpb@unam.mx

ABSTRACT
In this work, two different formulations from bone cements based in alpha tricalcic phosphate were compared. The first one corresponds to one commercial formula of bone cement, the second consists in alpha tricalcic phosphate (α-TCP), hydroxyapatite (HA), and zirconium oxide particles like reinforcement material.

The characterization of the cements consisted on physical tests of cohesion time, setting temperature and injectability of the paste, mechanical tests of compressive strength and adhesion force of the cement, finally X-rays diffraction (DRX) and electronic microscopy (SEM) to test the chemical and structural evolution of the setting reaction.

The second formulation shows better mechanical and chemical proprieties than the cement bone with commercial formulation. In addition, it presents an endothermic reaction that has the vantage of no injure the adjacent tissue.

INTRODUCTION
To anchor the implant with bone, the most useful cement is the PMMA, nevertheless this material has an exothermic reaction causing problems to the adjacent tissue [1]. An alternative are the cements of calcium phosphate [2, 3]. They have an endothermic reaction and the final product of their reactions in the body is hydroxyapatite, they reaching thus a better union between the implant and the bone [4,5].

However, it is desirable that the cement paste indicate an increase in its mechanical properties in the shortest possible time without affecting its injectability and manipulation. The hardening of the cement paste is related to the establishment of reaction and the final structure of cement.

The temperature, workability, cohesion time, and setting time were measured in the cement pastes. In addition, mechanical tests of compression and adhesion were made to the hard cements at different times since their preparation. To determine their crystalline phases, X-ray diffraction was used. Their microstructures were observed by SEM. In this work, a calcium phosphate cement (CPC) was developed and compared with a commercial formulation of a CPC.

MATERIALS AND METHODS
Cement reagents
The commercial CPC was named A; its reported composition for the solid phase was α-TCP 61% wt, $CaHPO_4$ 26% wt, $CaCo_3$ 10% wt and HA 3% wt, and the liquid phase was of a solution of Na_2HPO_4 2% wt. The liquid/powder rate (L/P) was 0.36 [6].

The reagents of CPC developed in this work and named B was: α-TCP 90% wt, HA 1% wt, and ZrO_2 9% wt. The liquid/powder rate (L/P) was a solution of Na_2HPO_4 at 0.6 M and a solution of $CaCl_2$ at 0.1 M, the L/P rate was 0.55.

The α-TCP and HA were synthesized in lab, whereas $CaCO_3$ and $CaHPO_4$ were obtained commercially from Backer and ZrO_2 was obtained from Riedel-of Haën.

39

HA was synthesized by hydrothermal methods, it consists of the reaction at room temperature of a solution 0.6 M of calcium hydroxide with a solution 0.3 M of phosphoric acid. The chemical reaction is:

$$10\ Ca(OH)_2 + 6\ H_3(PO_4) \rightarrow Ca_{10}(PO_4)_6(OH)_2 + 18\ H_2O$$

The preparation of α-TCP was carried out using solid state reaction, mixing stequiometric amounts of $CaCO_3$ and $CaH_4O_8P_2 \cdot H_2O$, the chemical reaction is:

$$2CaCO_3 + CaH_4O_8P_2 \cdot H_2O \longrightarrow \alpha\text{-}Ca_3(PO_4)_2 + 2CO_2 + 3H_2O$$

The reagents were placed in a Pt crucible at 1300°C during 12h and then quenched at room temperature on a cupper plate.

The addition of ZrO_2 was to improve the mechanical properties [7].

The CPC properties are determined largely by size of particles, thus, all components of cements were reduced on average at 20 micrometers, using an agate balls mill.

Characterization of CPC

The preparation of the cement was made as follows: the solid phase was added to liquid phase and then, both phases were mixed during one minute. The cement paste was deposited in a teflon tubes of 10 mm of diameter with a height of 13mm in order to obtain cylindrical samples of CPC.

To determine the setting temperature was used a digital thermometer, were made measuring data during 3 h every 30 s. Setting time was determined using the Gilmore needles.

Cohesion time was measured using cylindrical tubes of 10 mm diameter and 5 mm height; these were submerged in a SBF at 37 °C and 7.4 pH [8, 9].

Injectability was determined measuring the maximum effort to extrude 1.5 cc of cement paste, through a syringe with capacity of 3 cc without needle [10]. The test was made in a universal machine Instron 5500R with a speed of 50 mm/min.

Mechanical characterization

Twenty four cylindrical samples with 10 mm height and 5 mm diameter for each of cement were prepared to measure compressive strength. After cohesion time, the samples were placed in contact with a simulated physiological solution (SBF). After of 1, 3, 7 and 30 days, the compressions tested to six samples were made for each of cement, with a universal machine Instron 5500R at 1 mm/min [11, 12].

In order to measure the adhesion force, were used 4 nails of stainless steel AISI 316L with 4 mm diameter and 30 mm height with different finish: polished mirror, sand bath (sand blast), polished with sandpaper 80 and rough finished. These nails were placed inner bone, in a hole of 5 mm diameter and 10 mm height. The space between the nails and bone was filled up with cement paste. These samples were submerged in SBF. The adhesion force was measured trough the force to draw the nails [7, 13], at 1, 3, 7 and 30 days after preparation, for this was used a universal machine Instron 5500R, with velocity of 1 mm/min [11], see Figure 1.

Evolution chemical and structural

To determine the chemical evolution of the CPC, were taken DRX spectra at 1, 3, 7 and 30 days, after their preparation, using a Siemens D 5000, the changes in HA precipitated was considered. To study the structural evolution using SEM [12], was used a Cambridge Stereoscan 440 in these same samples (1, 3, 7 and 30 days).

RESULTS AND DISCUSSION
Mechanical characterization

Table 1 shows cohesion time for the cements. The cement B shows cohesion from the first minute after its preparation, reason for which can be placed in contact with liquid substances quickly. Meanwhile, the cohesion time of A cement is 6 h. Maybe this cement needs a promoter that is not reported in its technical report.

Table 1. Cohesion times of A and B cements.

	Time of cohesion [min]	
Cement	Experimental	Reported
A	360	< 5
B	1	-

Table 2 shows the setting times for the cements. The final setting time for B cement is superior to A cement. The final setting time reported for A cement is 18 min, maybe it is reach using a promoter.

Table 2. Setting times of A and B cements

	Initial setting time		Final setting time	
	Experimental	Reported *	Experimental	Reported *
Cement	[h]	[min]	[h]	[min]
A	2:00	~ 6	3:40	~ 18
B	1:30	-	9:30	-

* Reported by Calcibon® of Merck

The setting temperatures of the samples were measured at different times, considering the change in temperature as $\Delta T = T - T_{room}$, the setting reaction was endothermic. It was found that ΔT minimum of A cement was $-4.6°C$, meanswhile to B cement was $-4.2°C$ and to C cement was $-3.4°C$, in all cases the minimum was presented at 20 min after preparation, see Figure 1.

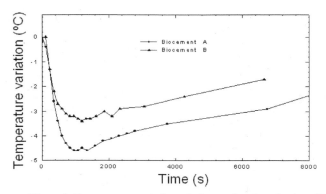

Figure 1. Temperature variations during setting time for A and B cements.

The cement is considered workable if it is suitable for injection. Supposing that the water is 100% injectable, and then a fluid that requires a greater effort to be injected will have a smaller value. Table 3 shows maximum effort applied to the cements and their injectability respect to the water. The behavior of A and B cements is very similar, showing an average injectability of 8.77% for A, and 8.05% for B. In B cement, the presence of the ZrO_2 and a greater presence of α-TCP influence on the injectability.

Table 3. Workability of the cements.

Sample	Maximum strength applied (MPa)	Injectability (%)
Water	0.33	100
Cement A	3.76	8.77
Cement B	4.05	8.14

Figure 2 shows the yield strength for the cements at different times during the setting and hardening processes. The fracture of cements was parallel to direction of load applied. The tendency of broken cement in compression is increased over time. During the first 7 days, the mechanical resistance increases, after this time stays constant, except for A cement that decays.

Compressive strength of A cement is less than B cement. It could be influenced for the microstructure and the zircon oxide presence at B cement. This is an indication of that the addition of zircon oxide particles improves adhesion but it reduces the compressive strength. Nevertheless, the mechanical proprieties of B cement were better than the A cement.

A larger roughness in polishing of the surface of the nails, the greater was the adhesion to the cement, the adhesion force increases with time in all cases, being B cement the most resistant; this behavior is showing in Figure 3.

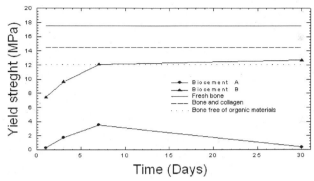

Figure 2. Yield strength of A and B cements at different time during setting and hardening processes.

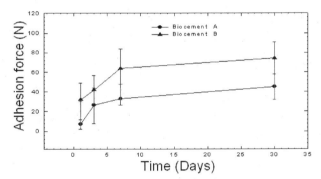

Figure 3. Adhesion force behavior versus time of A and B cements.

Evolution Chemical and Structural

Figures 4 and 5, show XRD patterns for A and B cements respectively at 0, 1, 3, 7 and 30 days after reaction. Whereas the cement A does not present any change with the time, B cement present variations with time. In the figures, the peaks corresponding to α-TCP and HA are indicated. At initial time, the reactants are forming the predominant phases, while HA peaks increase with time and the α-TCP peaks decreasing. After 7 days the mechanical properties of B cement reached the maximum, the peaks of zirconium oxide were constant during the reaction because this material is inert in this system.

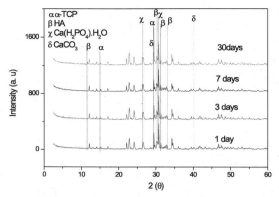

Figure 4. X ray diffraction pattern to A cement at different times of reaction

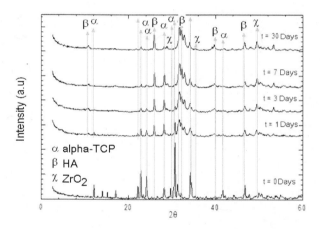

Figure 5. X ray diffraction pattern to B cement at different times of reaction.

Structural evolution

The microstructure study of these cements shows morphology as interconnected plates, they are flakes that seem to grow from a same center; if the flakes are seen from the side; they give the impression of needles distributed at random.

SEM images were obtained for each cement, at different times. In Figure 6, can be seen that A cement does not change in time. A cement shows agglomerated particles of the initial solid phase accordin with DRX pattern, where this cement not showed any morphologic changed.

Figure 6. Images by SEM at 10 000X of A and B cements at different times of reaction.

B cement presented, since the first day, an intercrossed plates morphology with a smooth center of a-TCP, whereas that the zircon oxide do not present a reaction with their surroundings. This smooth center disappears when the a-TCP begun to transformed in HA, after 3 days of immersion.

CONCLUSIONS

The B cement showed a satisfactory cohesion time and a good injectability. Its setting reaction is endothermic; it gives advantages over other cements. Also, B cement presents a greater adhesion than A cement, probably due to the zirconium oxide added. Their mechanical properties after seven days from initiate the setting are better compared with A cement.

The adhesion depends on the texture of the surface, being the better surface rough finish. The zirconium oxide added to B cement, increases considerably the adhesion of the cement forming a concrete.

The evolution of the chemical reaction and structural evolution of B cement are not affected by the addition of zirconium oxide particles.

ACKNOWLEDGEMENTS

To CONACYT by their economical support to Dr. H. H. Rodríguez and to DGAPA by economical support trough the project IN104008 and I thank to Dr. E. B Montufar especially for his great contribution in this work.

REFERENCES

[1]F. C. M. Driessens, J. A. Planell, F. J. Gil "Calcium Phosphate Bone Cements". Ency-clopedic HandBook of Biomaterial end Bioengineering Part B Application, **2**, Marcell Dekker, U.S.A, (1995), 855-877.

[2]E. M. Ooms, J. G. C. Wolke, J. P. C. M. Van der Waerden, J. A. Jansen "Use of Injectable Calcium-Phosphate Cement for the Fixation of Titanium Implants: an Experimental Study in Goats". J. Biomed. Mater. Res. (Appl. Biomater.), **66B**, (2003), 447-456.

[3]B. R. Constantz, I. C. Ison, M. T. Fulmer, R. D. Poser, S. T. Smith, Van Wagoner, J. Ross, S.A Goldstein, J. B. Jupiter, D. I. Rosenthal. "Skeletal Repair by *in situ* Formation of the Mineral Phase of Bone". Science, **276**, (1995), 1796-1799.

[4]E. Fernandez, F. J. Gil, M. P. Ginebra, F. C. M. Driessens, J. A. Planell "Calcium Phosphate Bone Cement for Clinical Applications Part I: Solution Chemistry". J.Mat.Sc.: Materials in Medicine, **10**, (1999), 169-176.

[5]E. Fernandez, F. J. Gil, M. P. Ginebra, F. C. M. Driessens, J. A. Planell. "Calcium Phosphate Bone Cement for Clinical Applications Part II: Precipitate Formation During Setting Reactions". J. Mat. Sc.: Materials in Medicine, **10** (1999), 177-183.

[6]www.biometmerck.com, Biomet Merck Biomaterials 2003.

[7]A. Quinto Hernández, Ma. Cristina Piña Barba. "Caracterización física y química de pastas de cementos óseos con ZrO_2". Rev. Mex. Fís. **49**, (2003), 123-131.

[8]E. Fernandez, M. G.Boltong, M. P. Ginebra, F. C. M. Driessens, O. Bermudez, J. A. Planell. "Development of a Method to Measure the Period of Swelling of Calcium Phosphate Cements". J. Mat. Sc.: Letters **15** (1996): 1004-1009.

[9]I. Khairoun, F. C. M. Driessens, M. G. Boltong, J. A. Planell, R. Wenz "Addition of Cohesion Promoters to Calcium Phosphate Cements". Biomaterials, **20**, (1999), 393-398.

[10]L. Roemhildt, T. D. Mcgee, S. D. Wagner "Novel Calcium Phosphate Composite Bone Cement: Strength and Bonding Properties". J. Mat. Sc.: Materials in Medicine, **14**, (2003), 137-141.

[11]E. Fernandez, F. J. Gil, S. M. Best, M. P. Ginebra, F. C. M. Driessens, J. A. Planell "Improvement of the Mechanical Properties of New Calcium Phosphate Bone Cement in the $CaHPO_4$-α-$Ca_3(PO_4)_2$ System: Compressive Strength and Microstructural Development". J. Biomed. Mater. Res. (Appl. Biomater.), **41**, (1998), 560-567.

[12]H.Yamamoto, S.Niwa, M.Hori, T.Hottori, K.Sawai, S.Aoki, M.Hirano, H.Takeuchi "Mechanical Strength of Calcium Phosphate Cement In Vivo and In Vitro". Biomaterials, **19**, (1998), 1587-1591.

[13]I. Khairoun, F. C. M. Driessens, M. G. Boltong, J. A. Planell, R. Wenz "Addition of Cohesion Promoters to Calcium Phosphate Cements". Biomaterials, **20**, (1999), 393-398.

Metallic Implant Materials

CHARACTERIZATION OF NEW NICKEL-TITANIUM WIRE FOR ROTARY ENDODONTIC
INSTRUMENTS

William A. Brantley and Jie Liu
College of Dentistry, Ohio State University
Columbus, OH, USA

William A.T. Clark, Libor Kovarik and Caesar Buie
Department of Materials Science and Engineering, Ohio State University
Columbus, OH, USA

Masahiro Iijima
School of Dentistry, Health Sciences University of Hokkaido
Ishikari-Tobetsu, Japan

Satish B. Alapati
College of Dental Medicine, Medical University of South Carolina
Charleston, SC, USA

William Ben Johnson
Sportswire LLC
Langley, OK, USA

ABSTRACT
    Rotary endodontic instruments are machined from pseudoelastic NiTi wires. Because
instruments occasionally fracture in root canals, there has been much effort to develop improved alloys.
Recently, rotary instruments have been introduced (Dentsply Tulsa Dental Specialties) that are
machined from a new wire (M-Wire) prepared by proprietary thermomechanical processing. M-Wire
has a higher ratio of tensile strength to upper pseudoelastic plateau stress than conventional NiTi wire
used to manufacture the rotary instruments, and instruments machined from M-Wire have superior
fatigue behavior. In this study, M-Wire and a conventional NiTi wire (Maillefer) were examined by
scanning transmission electron microscopy, optical microscopy, x-ray diffraction, and differential
scanning calorimetry to learn if differences in microstructures might account for observed differences
in mechanical properties. Results show that the conventional NiTi wire contains substantial austenite,
whereas there is a greater amount of martensite in M-Wire. M-Wire has coarser grains than the
conventional NiTi wire, and deformation bands and microtwinning provide strengthening. Evidence is
presented that differential scanning calorimetry does not detect the stable martensite in these wires.

INTRODUCTION
    The introduction of the nickel-titanium alloy to endodontics originated with the pioneering
study by Walia et al.,[1] which showed the much greater flexibility of NiTi for hand instrumentation
compared to the previously used austenitic stainless steel. Subsequently, rotary instruments used with
slow-speed dental handpieces, which permit rapid preparation of the root canal, were introduced and
are now in widespread clinical use. These instruments are machined from starting pseudoelastic
(termed *superelastic* in the dental literature) NiTi wire blanks,[2] and the inevitable surface defects from
the machining process have been implicated as a source of instrument fracture during endodontic
therapy.[3] While the incidence of clinical fracture is not great,[4] it is a source of anguish to patients and

has stimulated interest in new strategies, such as ion implantation[5] and electropolishing[6] for improving the fracture resistance of these instruments.

Recently, a proprietary procedure that involves extensive thermomechanical processing has been developed (Sportswire LLC), which yields NiTi wire (termed *M-Wire*) with greatly improved cyclic fatigue performance. M-Wire is characterized by a higher ratio of tensile strength to pseudoelastic upper plateau stress, compared to the NiTi wires that have been previously used for fabrication of rotary instruments. Differential scanning calorimetry has shown that these traditional NiTi rotary instruments are manufactured in the pseudoelastic condition, with the austenite-finish (A_f) temperature substantially below mouth temperature[7,8] and appear to remain in the pseudoelastic condition after simulated clinical use[9]. The objective of this study was to determine the origin of the improved mechanical properties for M-Wire, using standard metallurgical characterization techniques.

EXPERIMENTAL PROCEDURES

Two batches of M-Wire and one batch of conventional superelastic NiTi wire (Maillefer) were provided (Sportswire LLC). Wire segments had 5 cm length and 1 mm diameter.

Using a focused ion beam technique, foils were prepared from wire cross-sections for bright-field observations of microstructures at high magnification, using a scanning transmission electron microscope (Tecnai TF-20, Philips) operated at 200 kV. An optical microscope with digital image capture and a scanning electron microscope were used to examine microstructures at lower magnification. An aqueous etching solution containing hydrofluoric, nitric and acetic acids was employed to reveal the microstructures.

Micro-X-ray diffraction analyses (Rint-2000, Rigaku) were performed[10,11] to identify the phases in the wires, using Cu Kα radiation, tube voltage of 40 kV and current of 200 mA. The analysis sites had approximately 200 μm dimensions.

Phase transformation behavior in the wires was investigated with temperature-modulated differential scanning calorimetry.[12,13] For the TMDSC analyses (Q1000 and Q2000, TA Instruments), test specimens consisting of four wire segments having lengths of 4 – 5 mm were first cooled to −125°C, then heated to 100°C and cooled back to125°C. A 2°C/minute linear heating and cooling rate was used, and a sinusoidal thermal oscillation of 0.318°C amplitude and 60 second period was superimposed. This provided heating-only conditions during the heating cycle and cooling-only conditions during the cooling cycle. Enthalpy changes and onset temperatures for the transformations were determined with the software associated with the TMDSC apparatus.

RESULTS

The STEM observations revealed considerable differences between the microstructures of the conventional Maillefer pseudoelastic wire and M-Wire, as shown in Figures 1 and 2, and in Figures 3 and 4, respectively, for the wire cross-sections. The grain size for M-Wire was substantially larger than the grain size for the Maillefer wire. Triple-point junctions of grain boundaries and regions of microtwinning were prominent in the microstructure of M-Wire, indicative of the substantial mechanical deformation and extensive annealing performed during thermomechanical processing of this wire. Much less microtwinning was found in the microstructure of the conventional pseudoelastic wire. The STEM observations suggest that the Maillefer pseudoelastic wire is largely austenite with some martensite and perhaps R–phase, and that M-Wire is largely martensite. Regions of different contrast arise from varying grain orientations and localized mechanical deformation.

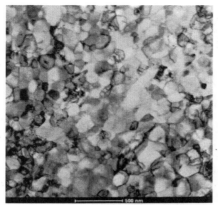

Figure 1. STEM microstructure of cross-section for
Maillefer pseudoelastic wire. Bar = 500 nm.

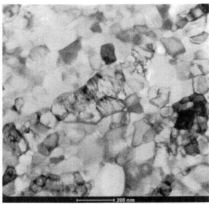

Figure 2. STEM microstructure of cross-section for Maillefer
wire at higher magnification, showing a region of substantial
microtwinning in the center of the photograph. Bar = 200 nm.

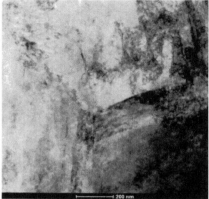

Figure 3. STEM microstructure of cross-section of M-Wire, showing a triple-point junction of three grains. Deformation bands and microtwinning are also evident. Bar = 200 nm.

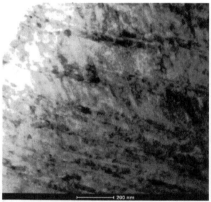

Figure 4. STEM microstructure of cross-section of M-Wire, showing another region with deformation bands and microtwinning. Bar = 200 nm.

Optical microscope photographs of polished and etched surfaces of conventional Maillefer pseudoelastic wire and M-Wire are shown in Figures 5 and 6, respectively. The microstructure of the Maillefer pseudoelastic wire contains wrinkled regions which are assumed to be martensite (or possibly R–phase) and some flattened regions which are assumed to be retained austenite. The microstructure of M-Wire appears to be largely martensite, with the classic lenticular appearance[14] of this type of structure.

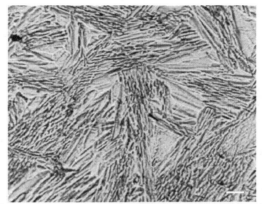

Figure 5. Microstructure of etched Maillefer pseudoelastic wire obtained with an optical microscope. Bar = 50 μm.

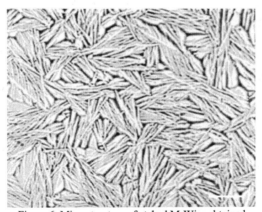

Figure 6. Microstructure of etched M-Wire obtained with an optical microscope. Bar = 10 μm.

Micro-XRD patterns at room temperature for conventional Maillefer pseudoelastic (superelastic) wire and the two batches of M-Wire are shown in Figure 7, with reflections from austenite, martensite and R–phase labeled. The peaks for Maillefer wire have been indexed to austenite, and the strong preferred orientation in this wire and M-Wire compared to the ICDD (International Centre for Diffraction Data) powder standard is evident. From the intensities of the martensite peaks, the first batch of M-Wire contains more martensite than the second batch. Two peaks for R–phase[15] are very near to both the 110 and the 211 peaks for austenite, and it is not possible to index these peaks in Figure 7 with certainty. Consequently, the reflections for both austenite and R–phase are shown.

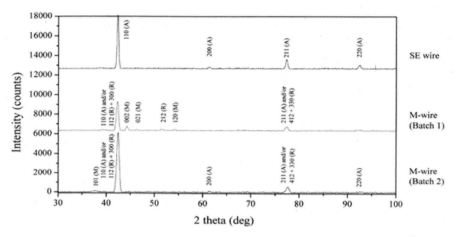

Figure 7. Micro-XRD patterns at room temperature for Maillefer superelastic (SE) wire (top) and two batches of M-Wire. Peaks are indexed to austenite (A), martensite (M) and R–phase (R).

The TMDSC total heat flow curves for the heating cycles (lower plots) and cooling cycles (upper plots) of the two batches of M-Wire are shown in Figure 8. The plots have been intentionally presented with slanting baselines for one batch and horizontal baselines for the other batch. The two endothermic peaks on the heating curves correspond to the transformation from martensite to R–phase at lower temperatures, followed by the transformation from R–phase to austenite at higher temperatures. The single exothermic peak on the cooling curves may correspond to the direct transformation from austenite to martensite, but it is possible that there is a two-step transformation involving the intermediate R–phase that cannot be resolved on this broad peak. The TMDSC plots for both batches of M-Wire were very similar, and the A_f temperature at which complete transformation to austenite occurred on heating was approximately 45° to 50°C. Lower values of A_f temperature are obtained from the intersection of a tangent line and the adjacent baseline,[16] and higher values of A_f temperature are obtained by a more conservative approach[7] where the curve returns to the baseline. At room temperature (approximately 25°C) the heating TMDSC plots suggest that M-Wire is largely a mixture of R–phase and austenite, with different proportions for the two batches, and the relative amount of austenite is higher at 37°C. For the Maillefer conventional pseudoelastic wire, the A_f temperature was approximately 10°or 15°C, depending upon whether this temperature was determined by the tangent line approach or the return of the heating curve to the baseline, respectively. Consequently, from the TMDSC heating curve the Maillefer pseudoelastic wire should be entirely austenite at room temperature, in agreement with the indexed Micro-XRD pattern in Figure 7.

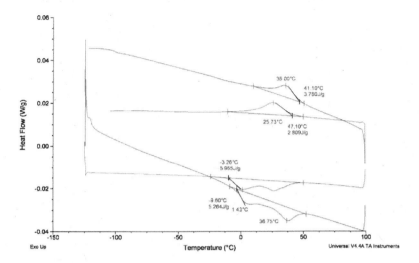

Figure 8. TMDSC total heat flow for two batches of M-Wire, which are distinguished by slanted and horizontal baselines. Lower curves are for heating cycle and upper curves are for cooling cycle. Construction lines are used by computer software to determine onset temperatures and enthalpy changes for phase transformations.

DISCUSSION

The STEM images in Figures 3 and 4 show that the microstructure of M-Wire presents evidence of substantial mechanical deformation during processing, with numerous deformation bands and considerable microtwinning. The much coarser grains and the triple-point junctions, compared to the microstructure of conventional Maillefer pseudoelastic wire in Figures 1 and 2, arise from extensive annealing during the proprietary thermomechanical processing. These strengthening mechanisms in M-Wire dominate the potential decrease in strength that would be expected from the much coarser grain size. The optical microscope images in Figures 5 and 6 show that M-Wire has a martensitic structure and that there is evidently substantial martensite in the Maillefer conventional pseudoelastic wire, as suggested by the STEM images in Figures 1 and 2. SEM photomicrographs of the microstructures were consistent with the optical microscope images.

There might appear to be some contradiction between the STEM and optical microscope results and the TMDSC analyses. TMDSC analyses (not shown) suggest that there should be minimal martensite in the Maillefer pseudoelastic wire at room temperature, which is much higher than the A_f temperature of $10° - 15°C$. Figure 8 shows that M-Wire should be a mixture of R–phase and austenite at room temperature, with minimal martensite present. The likely explanation is that the thermomechanical processing of Maillefer pseudoelastic wire and M-Wire creates substantial amounts of stable martensite that do not transform to austenite at the temperatures used for the present TMDSC analyses. Hence this martensite would not be readily detectable by these analyses. Support for this conjecture is given by the similar enthalpy changes for the overall transformation between martensite and austenite found for the Maillefer wire and M-Wire, which are much lower than those found for pseudoelastic orthodontic wires[12,13] and by the very low enthalpy changes found for the Classic Nitinol nonsuperelastic orthodontic wire[13] which has a largely stable work-hardened martensite structure.[17]

The Micro-XRD patterns in Figure 7 suggest that there are differences in the relative proportion of martensite in the microstructures of the two batches of M-Wire. Rotary endodontic instruments fabricated from M-Wire have recently become commercially available (Dentsply Tulsa Dental Specialties), and private communications with personnel at this company indicate that there have been changes in the original processing procedure developed at Sportswire LLC. Studies are currently in progress in our laboratories to understand the physical metallurgy[18] and relevant mechanical properties of the M-Wire instruments, using the techniques described in this article. Future transmission electron microscopy studies are planned with the use of sufficiently thin foils to obtain electron diffraction patterns for unambiguous determination of NiTi phases in as-received and clinically used instruments.

CONCLUSIONS

Nickel-titanium rotary endodontic instruments have complex microstructures that can contain martensite, austenite and R–phase. The superior mechanical properties of M-Wire are due to alloy strengthening mechanisms that are the result of an extensive thermomechanical processing procedure whose details are proprietary. Full understanding of the structure-property relationships for these rotary instruments requires the use of several complementary techniques, such as transmission electron microscopy, x-ray diffraction, differential scanning calorimetry, and metallographic observations of etched microstructures. Evidently, the manufacturing process for these instruments results in the formation of substantial amounts of stable work-hardened martensite that do not transform over the temperature range for the thermal analyses in the present study. Additional measurements of hardness using both the Vickers microindentation procedure and a nanoindenter can serve as probes to provide details about variations in work-hardening for as-received and clinically used instruments at different positions along the shafts and at different areas in the cross-sections.

REFERENCES
[1]H. Walia, W.A. Brantley, and H. Gerstein, An Initial Investigation of the Bending and Torsional Properties of Nitinol Root Canal Files, J. Endod., 14, 346-51 (1988).
[2]S.A. Thompson, An Overview of Nickel–Titanium Alloys Used in Dentistry, Int. Endod. J., 33, 297-310 (2000).
[3]S.B. Alapati, W.A. Brantley, T.A. Svec, J.M. Powers, J.M. Nusstein, and G.S. Daehn, SEM Observations of Nickel-Titanium Rotary Endodontic Instruments that Fractured During Clinical Use, J. Endod., 31, 40-3 (2005).
[4]P. Parashos and H.H. Messer, Rotary NiTi Instrument Fracture and Its Consequences, J. Endod., 32, 1031-43 (2006).
[5]T.B. Bui, J.C. Mitchell, and J.C. Baumgartner, Effect of Electropolishing ProFile Nickel-Titanium Rotary Instruments on Cyclic Fatigue Resistance, Torsional Resistance, and Cutting Efficiency, J. Endod., 34, 190-3 (2008).
[6]M.E. Anderson, J.W. Price, and P. Parashos, Fracture Resistance of Electropolished Rotary Nickel-Titanium Endodontic Instruments, J. Endod., 33, 1212-6 (2007).
[7]W.A. Brantley, T.A. Svec, M. Iijima, J.M. Powers, and T.H. Grentzer, Differential Scanning Calorimetric Studies of Nickel Titanium Rotary Endodontic Instruments, J. Endod., 28, 567-72 (2002).

[8]G. Kuhn and L. Jordan, Fatigue and Mechanical Properties of Nickel-Titanium Endodontic Instruments, J. Endod., **28**, 716-20 (2002).

[9]W.A. Brantley, T.A. Svec, M. Iijima, J.M. Powers, and T.H. Grentzer, Differential Scanning Calorimetric Studies of Nickel-Titanium Rotary Endodontic Instruments after Simulated Clinical Use, J. Endod., **28**, 774-8 (2002).

[10]M Iijima, H. Ohno, I. Kawashima, K. Endo, W.A. Brantley, and I. Mizoguchi, Micro X-ray Diffraction Study of Superelastic Nickel-Titanium Orthodontic Wires at Different Temperatures and Stresses, Biomaterials, **23**, 1769-74 (2002).

[11]M. Iijima, W.A. Brantley, I. Kawashima, H. Ohno, W. Guo, Y. Yonekura, and I. Mizoguchi, Micro-X-ray Diffraction Observation of Nickel-Titanium Orthodontic Wires in Simulated Oral Environment, Biomaterials, **25**,171-6 (2004).

[12]W.A. Brantley, M. Iijima, and T.H. Grentzer. Temperature-Modulated DSC Study of Phase Transformations in Nickel-Titanium Orthodontic Wires, Thermochimica Acta, **392-3**, 329-37 (2002).

[13]W.A. Brantley, M. Iijima, and T.H. Grentzer, Temperature-Modulated DSC Provides New Insight about Nickel-Titanium Wire Transformations, Am. J. Orthod. Dentofacial Orthop., **124**, 387-94 (2003).

[14]R.E. Reed-Hill and R. Abbaschian, Physical Metallurgy Principles (3rd ed), Boston: PWS Publishing, pp. 538-87 (1994).

[15]G. Riva, M. Vanelli, and T. Airoldi, A New Calibration Method for the X-ray Powder Diffraction Study of Shape Memory Alloys, Phys. Stat. Sol. A **148**, 363-72 (1995).

[16]International Organization for Standardization, ISO Standard 15841, Dentistry—Wires for Use in Orthodontics (December 2006).

[17]T.G. Bradley, W.A. Brantley, and B.M. Culbertson, Differential Scanning Calorimetry (DSC) Analyses of Superelastic and Nonsuperelastic Nickel-Titanium Orthodontic Wires, Am. J. Orthod. Dentofacial Orthop., **109**, 589-97 (1996).

[18]K. Otsuka and X. Ren. Physical metallurgy of Ti–Ni-based shape memory alloys. Progress Mater. Sci., **50**, 511-678 (2005).

EFFECT OF COLD WORK ON THE BEHAVIOR OF NiTi SHAPE MEMORY ALLOY

Mohamed Elwi Mitwally and Mahmoud Farag
Department of Mechanical Engineering
The American University in Cairo, Egypt

ABSTRACT

An annealed NiTi alloy (50.7 at% Ni) was subjected to cold rolling with reductions of 20%, 30% and 40%. The effect of cold rolling on the structure was investigated using X-ray diffraction, SEM and optical microscopy. The mechanical properties were measured using tension and nanohardness tests. Superelasticity was measured from the loading-unloading curves of the nanoindenter.

The results show that increasing cold rolling causes progressive increase in Martensite in the alloy structure. Increasing cold rolling was also found to cause an increase in superelasticity, tensile strength and hardness but reduces ductility. Most of the properties tested (P) were found to vary with the amount of cold rolling (CR) according to a quadratic polynomial relationship:

$$P = a\,(CR)^2 + b\,(CR) + c$$

where a, b, and c are constants.

The use of the above relationship in predicting the behavior of the NiTi alloy under different cold rolling conditions is discussed.

INTRODUCTION

Shape memory alloys are characterized by their ability to recover high values of strain when heated, after being deformed at a low temperature (shape memory effect) and their ability to recover high values of strain upon removal of mechanical load (superelasticity). Superelasticity has also been defined to be the ability of the material to store energy [1]. Such behavior is attributed to the phase transformation that Shape memory alloys undergo from a low temperature phase (Martensite) to a high temperature phase (Austenite) upon the application of thermal or mechanical load.

The main requirement for achieving superelasticity is that the stress needed to induce Martensite should be lower than that to cause dislocation slip otherwise the material would show no unique behavior. Raising the stress needed for slip can be done either by precipitation hardening or by introducing more dislocations [2]. Earlier work has shown that cold working of NiTi alloys can increase their superelasticity and causes them to exhibit a plateau when subjected to tension in the rolling direction only [3, 4]. It has also been shown that cold work increases the number of dislocations in the material which in turn raises the strength and hardness but decreases the ductility [5]. Cold rolling has been used by several researchers to impart cold work to NiTi [6-13]. Hosoda et al. [6] found that hardness increases as cold rolling increases when dealing with austenitic NiTi alloys. Hardness of martensitic NiTi also increases with cold rolling percentage [7,9].

The aim of this study is to examine the effect of cold work on the behavior of NiTi shape memory alloy.

MATERIALS AND EXPERIMENTAL PROCEDURE

The NiTi alloy used in this work has 55.8at% Ni content by weight (50.7at%) with the rest being Ti. It was supplied by EUROFLEX GmbH, in the form of a hot rolled rod that was annealed at 800-850°C. According to the supplier, the transformation temperatures are M_f = -53.7 °C, M_s= -38.9

$^{\circ}$C, A_s = -33.5 $^{\circ}$C and A_f ranges between 5 to 18 $^{\circ}$C. Strips were cut parallel to the rod axis using wire electric discharge machining (EDM) and cold rolled by thickness reductions of 20, 30 and 40%.

The material was then tested to determine the effect of cold work on the structure (using X ray diffraction and microscopy analysis), mechanical properties (using nanohardness and tensile tests), and superelasticity (using the ratio between the energy stored during loading and the energy released during unloading in a nanohardness test). Following is a description of the experimental techniques used:

XRD analysis

X ray diffraction was performed to determine the phases present using a Philips XRD system equipped with a CuKα source. Data refinement, fitting and background elimination was done using Peakfit from Systat Software Inc.

Further analysis was done using the "external standard method" to have some quantitative analysis for phase fractions present in each sample according to Cullity et al. [14]. Full widths at half heights of the 110 Austenite peak were used to qualitatively describe the change in the crystal size of the Austenite, as it was the Austenite peak with highest relative intensity.

Microscopy analysis

Samples were prepared for microscopy analysis using grinding, polishing and etching using a solution of HF, HNO_3 and H_2O with volume ratios of 1:4:5 respectively [10, 12]. A LEO SUPRA 55 SEM operated at 10-20 KV was used for microstructural examination. The line intercept method was used to calculate the grain size.

Tension testing

Tensile samples were cut parallel to the rolling direction using wire EDM machining. The samples conformed to ASTM standards with a gauge length 6 mm and width 2 mm. Two samples were used for each condition and a third was used in case of discrepancy between the first two samples. Tests were carried at a constant strain rate of 1.67×10^{-4} /s using an INSTRON 5569 machine with a 50kN load cell. Strain rate control was used to insure that a plateau would be observed at least for the as received sample since load control would eliminate the plateau [4]. The low value for strain rate was also used to minimize temperature changes that might occur during the test [15, 16].

Nanohardness measurements

Nanohardness measurements were carried out using an MTS XP nanoindenter on polished surfaces to insure low surface roughness that might lead to errors in depth calculations [17]. A diamond Berkovich tip was used at a depth of 2000 nm in all samples to eliminate the effect of depth variation. Nanohardness was taken as the maximum force divided by the area of contact.

Superelasticity analysis

In the present work, superelasticity is defined as the ratio between the energy released from the material while unloading (E2) to the energy stored while loading (E1), which is the approach that has been used in several studies [8, 18, 19]. The MTS XP nanoindenter was used to obtain load-unload curves. A tungsten carbide 400 microns diameter spherical tip was used to measure the superelasticity. A 500 mN load was used to minimize surface roughness effects, which decrease with the increase of applied load and increase with the decrease in tip diameter [17]. Polynomial regression was used to fit equations for both the loading and unloading curves as they are supposed to a follow a polynomial relation [17]. Areas under loading and unloading scurves were computed to measure superelasticity.

RESULTS & DISCUSSION

XRD results

Fig. 1 shows XRD charts for as received and as cold rolled NiTi samples after data refinement. Only three Austenite peaks were observed in all charts, the 110, 200 and 211 peaks. Several peaks were observed for Martensite with the 002 and 1-11 being the most significant, having appeared in all the charts including the as received sample. Only these two peaks are shown here for clarity.

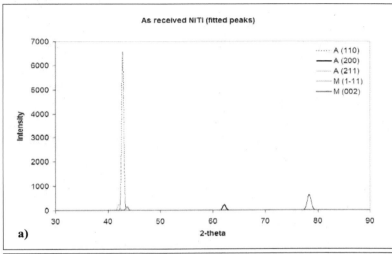

a)

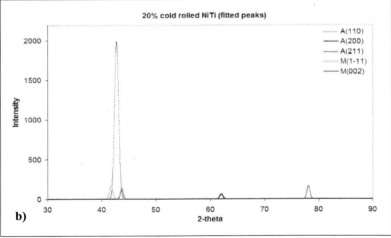

b)

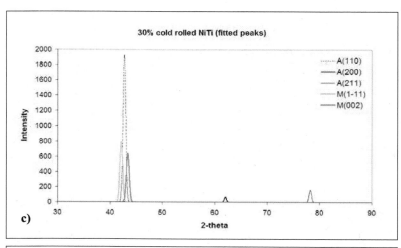

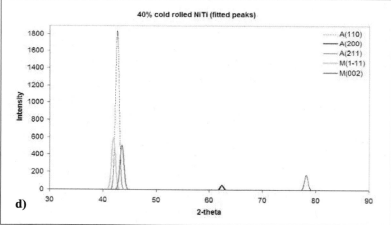

Fig. 1. XRD charts for (a) as received, (b) 20%, (c) 30% and (d) 40% cold rolled samples after fitting.

Fig. 1 shows that the 110 Austenite peak broadens with cold rolling. The full width at half maximum intensity (FWHM) of the 110 Austenite peak vs. cold rolling is shown in Fig. 2. The broadening of the 110 peak was observed by Tsuchiya et al. [11]. Such behavior has been attributed to crystal refinement and cold work defects that cause strain nonuniformity which lead to peak broadening [14].

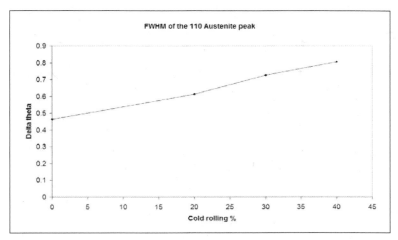

Fig. 2. Peak broadening of the 110 Austenite peak with cold rolling.

Results from the external standard method in Fig. 3 show a clear increase in Martensite volume fraction, which reaches about 60% in the 40% cold rolled sample.

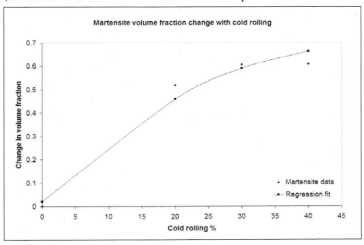

Fig. 3. Martensite volume change vs. cold rolling percentage.

The relation between the change in Martensite volume fraction and cold rolling percentage was found to follow the quadratic relation:

$$M = -0.00029 \, (CR)^2 + 0.0277 \, (CR) + 0.0209 \qquad (1)$$

with an R value of 0.986, where M is the change in Martensite volume fraction and CR is the cold rolling percentage. Eq. (1) predicts that Martensite will continue to increase till about 48% of cold rolling and will decrease afterwards. This means that Austenite is transformed into stress induced Martensite up to a point where any further increase in cold rolling would cause a decrease in Martensite volume fraction. This has also been observed by Nakayama et al. [10] who have explained this behavior to be due to the increase in dislocation density at some cold rolling stage that would stabilize the Austenite phase at the expense of Martensite.

Microstructure analysis
    SEM images taken for the transverse sections of as received and cold rolled NiTi are presented in Fig. 4. The grain size of the as received NiTi was about 26μm as measured by the line intercept method.

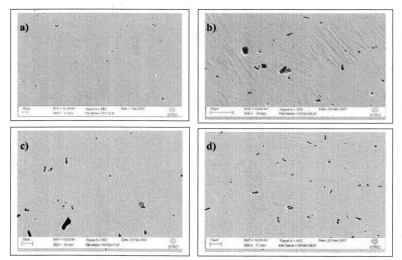

Fig. 4. SEM images for (a) as received NiTi, (b) 20% cold rolled NiTi, (c) 30% cold rolled NiTi & (d) 40% cold rolled NiTi

Fig. 4 shows that the grains in the as received NiTi are equiaxed and that there are few Martensite variants (groups) that can be identified within some grains, while other grains are nearly free of any Martensite. The Martensite phase is seen to increase with increasing cold rolling percent, which agrees with the XRD results.

Tensile tests

Fig. 5 shows representative tensile stress strain curves of the as received material and the cold rolled samples. The tensile stress-strain curves show that the plateau region is present only in the as received samples and occurred at a stress range of 425 to 459 MPa and results in strains in the range of 7.5 to 9.2%. This is in agreement with the results of Lin et al. [8], who reported a complete disappearance of plateau in cold rolled NiTi after 10% cold rolling.

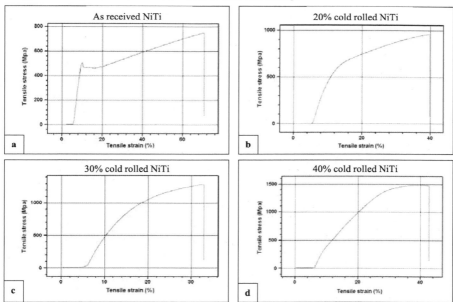

Fig. 5. Tensile stress strain curves for (a) as received NiTi (b) the 20%, (c) the 30% and (d) the 40% cold rolled sample.

Fracture stress and strain are plotted against cold rolling in Fig. 6 and Fig. 7 respectively. Fig. 6 and Fig. 7 show that the maximum stress in the tension test increases and the total strain decreases with increasing cold rolling percent. Both the relations were found to follow the quadratic relations:

$$\sigma_f = 0.3074 \, (CR)^2 + 6.7743 \, (CR) + 754.59, \, R=0.983 \quad (2)$$

$$\varepsilon_f = 0.037 \, (CR)^2 - 2.586 \, (CR) + 71.554, \, R=0.968 \quad (3)$$

where σ_f and ε_f in equations 2 and 3 represents the fracture stress and fracture strain respectively with CR being the percentage of thickness reduction. Both Eq. (2) and Eq. (3) show that as cold rolling increases strength increase but ductility decreases which can be attributed to the strengthening effect of the rolling process which causes an increase in the dislocation density as the thickness reduction percentage increases. The slight increase in the fracture strain at 40% cold rolling can be attributed to the increase in temperature during the rolling process at this high reduction. Such increase in temperature could lead to a recovery stage that is characterized by some increase in ductility with no decrease in strength.

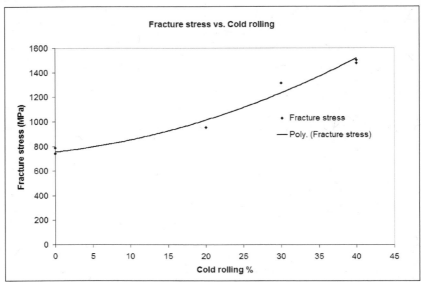

Fig. 6. Fracture stress vs. cold rolling for NiTi samples under tension testing.

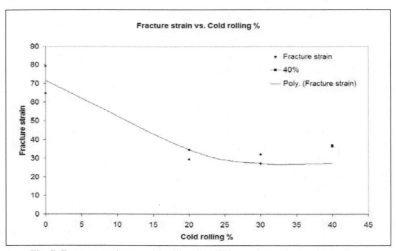

Fig. 7. Fracture strain vs. cold rolling for NiTi samples under tension tesing.

Hardness results

Fig. 8 shows that the nanohardness increased with the increase in cold rolling percentage. Such an increase in hardness with cold rolling has been observed with Vickers hardness of austenitic or martensitic NiTi alloys and is attributed to the dislocations produced due to cold rolling [7, 9, 10].

The increase in hardness with thickness reduction was found to follow the quadratic relation:

$$H = 0.000584 \ (CR)^2 - 0.00292 \ (CR) + 3.865, \ R = 0.996 \tag{4}$$

where H represents hardness values and CR represents the thickness reduction as defined before. Eq. (4) shows that hardness continues to increase progressively with cold rolling.

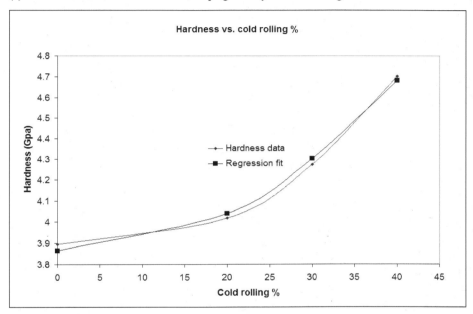

Fig. 8. Hardness of as received and cold rolled NiTi samples as a function of thickness reduction.

Superelasticity results

Average superelasticity values, defined as E2/E1, are shown in Fig. 9 with the dashed line representing the superelasticity value for as received NiTi.

Fig. 9 shows that superelasticity increases with cold rolling with the as received NiTi showing the highest superelasticity among all samples. As received annealed NiTi is known to have higher superelasticity than cold worked NiTi which has been associated with recoverable strains of 2-4% [3, 11]. The increase in superelasticity with cold rolling follows the relation:

$$SE = -0.0001463 \ (CR)^2 + 0.0111413 \ (CR) + 0.7127245, \ R=1 \tag{5}$$

where SE represents average superelasticity and CR represents the cold rolling percentage. As can be predicted from Eq. (5), superelasticity of cold rolled NiTi will never reach that of as received NiTi with any further increase in cold rolling. Actually, the highest value of superelasticity that could be reached is supposed to occur at 38% cold rolling according to Eq. (5). That value is about 0.925 as compared to 0.944 for the as received NiTi. The cold rolling value where maximum superelasticity occurred with cold rolled NiTi agrees with what has been observed in literature to be a value between 30-40% where optimum performance is supposed to occur [20].

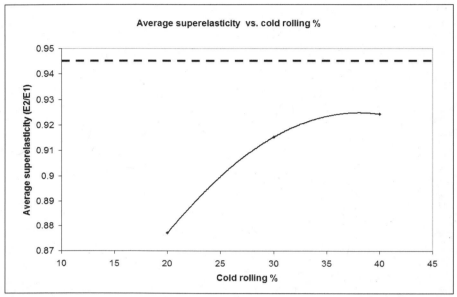

Fig. 9. Average superelasticity values for the as received and as cold rolled samples measured by the spherical tip

Having the highest Austenite volume fraction could be the reason why as received NiTi showed highest superelasticity in the present study since Austenite is the phase responsible for the superelastic behavior.

The increase in superelasticity for cold rolled martensitic NiTi has been observed in literature and has been attributed to the strengthening effect of cold rolling that makes dislocation slip more difficult to occur than the stress induced Martensite transformation. This leads to the material withstanding higher values of stress without slip while Martensite is being stress induced thus causing superelasticity to occur up to higher values of stress [8, 12, 21].

CONCLUSION

In the present study, an annealed 50.7at% Ni NiTi alloy was subjected to cold rolling at reductions of 20%, 30% and 40%. Structure changes and mechanical properties were investigated for as received, and cold rolled NiTi samples.

1. XRD charts show that some Martensite exists in the as received annealed material. The Martensite volume fraction increases with cold rolling at the expense of Austenite. The increase in Martensite volume fraction with cold rolling followed the quadratic relation in Eq. (1).

2. Stress strain curves of as received NiTi showed the traditional plateau associated with superelasticity. Cold rolled NiTi samples showed no superelasticity plateau.
3. Strength of cold rolled samples increased compared to that of as received NiTi while ductility generally decreased due to the strengthening effect caused by cold rolling. The increase in strength and the decrease in ductility with cold rolling both showed quadratic behavior according equations 2 and 3.
4. Hardness values also increased due to the strengthening effect of cold rolling according to Eq. (4).
5. Superelasticity as measured by the nanoindenter showed an increase with cold rolling according to Eq. (5).
6. As received NiTi showed higher superelasticity than the cold rolled samples. After a large drop at 20% due to a decrease in Austenite, superelasticity then increased with cold rolling. Such increase is attributed to the strengthening effect of cold rolling.

Using cold rolling alone as a technique to improve the mechanical performance of NiTi SMAs has had a significant impact. A significant gain in strength and hardness was achieved without a significant decrease in superelasticity occurring. The change in properties with cold rolling was modeled to follow quadratic equations of the form:

$$P = a\,(CR)^2 + b\,(CR) + c$$

which were further used to predict any change in performance with increasing cold rolling percentage.

ACKNOWLEDGEMENTS
This research was funded by the Youssef Jameel Science & Technology Research Center (STRC) at the American University in Cairo (AUC). The authors would like to thank Dr. Hanady Salem, Dr. Mohamed Gaafar, Omar Mortagy, Ahmed Nagy and Hanady Hussein for their assistance.

REFERENCES
[1]T.W. Duerig and A.R. Pelton in: R. Boyer, G. Welsch and E.W. Collings (Eds.), Materials Properties Handbook: Titanium Alloys, ASM International, 1994, pp. 1035-1048.
[2]S. Miyazaki in: T.W. Duerig, K.N. Melton, D. Stoeckel and C.M. Wayman (Eds.), Engineering aspects of shape memory alloys, Butterworth-Heinmann, England, 1990, pp. 394-413.
[3]G.R. Zadno and T.W. Duerig in: T.W. Duerig, K.N. Melton, D. Stoeckel and C.M. Wayman (Eds.), Engineering aspects of shape memory alloys, Butterworth-Heinmann, England, 1990, pp. 414-419.
[4]P. Sittner P, Y. Liu and V. Novak, On the origin of Luders-like deformation of NiTi shape memory alloys, Journal of the Mechanics and Physics of Solids, 53, 1719-1746 (2005).
[5]R. Reed and R. Abbaschian, Physical Metallurgy Principles, PWS-KENT publishing company, Boston, 1991.
[6]H. Hosoda, S. Miyazaki, K. Inoue, T. Fukui, K. Mizuuchi, Y. Mishima and T. Suzuki, Cold rolling of B2 intermetallics, Journal of Alloys and Compounds, 302, 266-273 (2000).
[7]H.C. Lin, S.K. Wu, T.S. Chou and H.P. Kao, The Effects of Cold Rolling on the Martensitic Transformation of an Equiatomic TiNi Alloy, Acta Metallurgica et Materialia, 39, 2069-80 (1991).
[8]H.C. Lin and S.K. Wu, The Tensile Behavior of a Cold-Rolled and Reverse- Transformed Equiatomic TiNi Alloy, Acta Metallurgica et Materialia, 42, 1623-30 (1994).
[9]H.C. Lin and S.K. Wu, Determination of Heat of Transformation in a Cold-Rolled Martensitic TiNi Alloy, Metallurgical Transactions A, 24, 293-299 (1993).
[10]H. Nakayama, K. Tsuchiya, Z.G. Liu, M. Umemoto, K. Morii and T. Shimizu, Process of nanocrystallization and partial amorphization by cold rolling in TiNi, Materials Transactions, 42, 1987-93 (2001).

[11]K. Tsuchiya, M. Inuzuka M, D. Tomus, A. Hosokawa, H. Nakayama, K. Morii, Y. Todaka and M. Umemoto, Martensitic transformation in nanostructured TiNi shape memory alloy formed via severe plastic deformation, *Materials Science and Engineering A*, **438-440**, 643-648 (2006).

[12]W.C. Crone, D. Wu and J.H. Perepezko, Pseudoelastic behavior of nickel-titanium melt-spun ribbon, *Materials Science and Engineering A*, **375-377**, 1177-81 (2004).

[13]J. Koike and D.M. Parkin, Crystal-to-amorphous transformation of NiTi induced by cold rolling, *Journal of Materials Research*, **5**, 1414-18 (1990).

[14]B.D. Cullity and S.R. Stock, Elements of X-ray Diffraction, Prentice-Hall inc, New Jersey, 2001.

[15]X. Huang and Y. Liu, Effect of annealing on the transformation behavior and superelasticity of NiTi shape memory alloy, *Scripta Materialia*, **45**, 153-160 (2001).

[16]J.M. McNaney, V. Imbeni, Y. Jung, P. Papadopolous and R.O. Ritchie, An experimental study of the superelastic effect in a shape-memory Nitinol alloy under biaxial loading, *Mechanics of Materials*, **35**, 969-986 (2003).

[17]A. Fischer-Cripps, Nanoindentation, Springer-Verlag, New York, 2002.

[18]R. Liu R and D.Y. Li, Indentation behavior of pseudoelastic TiNi alloy, *Scripta Materialia*, **41**, 691-696 (1999).

[19]W. Ni and Y. Cheng, Microscopic superelastic behavior of a nickel-titanium alloy under complex loading conditions, *Applied Physics Letters*, **82**, 2811-13 (2003).

[20]A. Serneels, Proceedings of the First European Conference on Shape Memory and Superelastic Technologies: SMST-99, SMST Europe, Lubbeek, Belgium, 6-23, (1999).

[21]D. Wu, W.C. Crone and J.H. Perepezko, Proceedings of the SEM Annual Conference on Experimental Mechanics, Milwaukee, WI, 1-4 (2002).

MICROSTRUCTURE AND MECHANICAL PROPERTIES OF Ti-6Al-4V FOR BIOMEDICAL AND RELATED APPLICATIONS INVOLVING RAPID-LAYER POWDER MANUFACTURING

L. E. Murr, S. M. Gaytan, S. A. Quinones, M. I. Lopez, A. Rodela, E. Y. Martinez, D. H. Hernandez, E. Martinez, D. A. Ramirez, F. Medina, R. B. Wicker
University of Texas at El Paso, El Paso, TX 79968 USA

ABSTRACT
    The microstructures and mechanical properties of rapid-layer manufactured Ti-6Al-4V specimens from precursor powder by electron beam melting (EBM) and selective laser melting (SLM) are compared with those characteristic of commercial wrought products of Ti-6Al-4V. The microstructures are characterized by optical metallography and scanning and transmission electron microscopy. The EBM built samples exhibited α (hcp) acicular platelet (basketweave or Widmanstätten) microstructure similar to wrought products while the SLM built samples exhibited primarily α' (hcp) martensitic microstructure. The EBM samples exhibited a UTS as high as 1.2 GPa and elongations in excess of 20% in contrast to the SLM samples which exhibited a UTS of 1.3 GPa and elongations of only 4%. Rockwell C-scale hardnesses ranged from 37 to 55 for the EBM samples indicative of a wide range of graded properties and complex monolithic and open (or porous) geometries which might be digitally manufactured for biomedical and aerospace applications in particular.

INTRODUCTION

    The fabrication of titanium and titanium alloy components for aerospace as well as biomedical applications involving millions of hip and knee replacements world-wide annually rely upon forgings or HIPed castings which require substantial rough and fine machining resulting in up to 80 percent material waste. Forgings of alpha/beta titanium alloys typically manufactured for the aerospace industry require a multi-stage, closed-die forging of wrought mill products which when combined with machining stages may exceed a 20:1 "buy-to-fly" ratio for large, complex airframe structural components.

    Solid freeform fabrication (SFF) or rapid prototyping (RP) comprises a broad class of single-step, waste-free processes to manufacture near net shape components using metal powders without dies and molds. In these methods of component production, metal or alloy powder is either fed at a controlled rate into the focal point of a laser or electron beam or spread (or raked) into a layer which is selectively sintered or melted and solidified by rastering of the beam using a CAD file to build components layer-by-layer. Processes based on laser deposition (LD) or direct laser fabrication (DLF) have involved laser engineered net shaping (LENS) and selective laser melting (SLM).[1-3] Correspondingly, electron beam melting (EBM) is an emerging direct digital manufacturing (DDM) or layered manufacturing (LM) process.[4,5] These SFF processes are particularly attractive for the manufacture of both aerospace and biomedical components because of lead-time reduction waste-free fabrication, cost-of-tooling reduction, prospects for microstructure-property manipulation,[6,7] and prospects for graded composition, structure, and properties; including porosity variations[8] and open structure geometries of varying density built upon fully dense, layer-built monoliths.

    The objective of the present work was to compare the microstructure and mechanical properties of simple Ti-6Al-4V component geometries fabricated from precursor powders

71

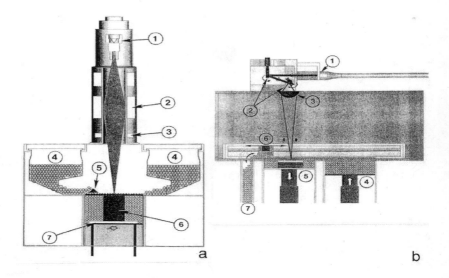

Figure 1. Schematic views of the EBM (a) and SLM (b) systems showing key components at numbered circles. (a) EBM system: (1) Electron gun assembly; (2) EB focusing lens; (3) EB deflection coils; (4) Powder cassettes; (5) Powder layer rake; (6) Test specimen builds; (7) Build table. (b) SLM system: (1) Laser; (2) Double rotating mirror system; (3) Beam focus lens; (4) Powder feed system; (5) Build platform; (6) Recoater or powder rake; (7) Powder recovery system. (From Murr, et al.[9] by permission from Elsevier, Ltd.)

utilizing EMB and SLM. These component structures and properties were also compared with wrought and cast Ti-6Al-4V products to provide a context for these SFF technologies.

EXPERIMENTAL

Figure 1 compares the commercial EBM and SLM systems utilized for this investigation. Figure 1(a), shows a schematic of the ARCAM EBM S400 system which builds ~100 μm thick layers by scanning the focused electron beam (EB) (shown by 2 and 3 numbered circles) in a vacuum at nominal scan rates of 10^3 mm/s to selectively melt the powder layer using a 3D-CAD system. The powder is continuously added from powder cassettes (4) to the top of the building part. The powder rake at (5) in Figure 1(a) moves laterally between the two cassettes (4) to evenly distribute the powder layer over the surface after each build layer is melted. The build (6) on the build table (7) moves down as layers are added. The electron beam is formed in the gun (1) in Figure 1(a) at 60 kV potential to develop a nominal energy density in the focused beam of $\sim 10^2 kW/cm^2$.

Figure 1(b) illustrates for comparison a schematic view of the EOSINT M270 SLM system which builds ~30 μm thick layers by scanning the focused laser (ytterbium fiber) beam (at 3 in the numbered circle) in a nitrogen environment at nominal scan rates of ~7 x 10^3 mm/s (nearly an order of magnitude greater than normal EBM) to melt the powder layer as in the EBM system; utilizing a similar power (or energy) density (~10^2 kW/cm^2). The powder cassette or feeder system (at 4 in Figure 1(b)) is raked over the build (5) as shown at 6 in Figure 1(b), and unused powder is recovered at (7).

The Ti-6Al-4V (Grade 5) powder, shown typically in Figure 2, had a mean diameter of ~30 μm, with a bimodal size distribution for the EBM powder, while the SLM Ti-6Al-4V powder had a 50% smaller mean diameter (20 μm), with a single-modal distribution which allowed for decreased layer dimensions. In Figure 2 the bimodal distribution is shown to arise in part by smaller atomized precursor powder particles sintered to larger ones. Figure 2(b) also shows an enlarged view of one of the bimodal powders (arrow in Figure 2(a)) exhibiting a generally equiaxed α-phase grain structure at the particle surface (arrow). As noted in Figure 3, especially Figure 3(b), this results by the acicular α plates intersecting the surface. Figure 3(a) shows a Vickers microindentation in the surface of a Ti-6Al-4V powder particle embedded in an epoxy mount and ground and polished with an alumina abrasive slurry, while Figure 3(b) shows similar embedded and polished powder particles etched with a solution consisting of 100 mL water, 5 mL HNO$_3$ and 2.5 mL HF to reveal the acicular α platelet microstructure characteristic of the atomized powder. These polishing and etching procedures were also utilized in revealing the microstructure of the EBM and SLM built components observed by both optical metallography and scanning electron microscopy (SEM). Optical metallography in this study utilized a Reichert MEF4A/M system while the SEM was a Hitachi S4800 field emission (FE) SEM. The EBM and SLM components were also cut into appropriate, thin coupons which were electropolished to electron transparent specimens for transmission electron microscopy (TEM) using a Stuerers Tenupol-5, Dual-Jet unit using a solution consisting of 0.9L methanol to which 50 mL H$_2$SO$_4$ was added. The solution was cooled to -10°C and the electropolishing performed at 5A in a voltage range from 15 to 25V. The TEM was a Hitachi H-8000 analytical electron microscope operated at 200 kV accelerating potential, employing a goniometer-tilt stage.

Vickers microindentation hardness (HV) measurements as illustrated in Figure 3(a) were made in a Shimadzu HMV-2000 microindentation hardness tester utilizing a 25 gf load on the Vickers diamond indenter. Rockwell C-scale hardness measurements (HRC) were also made on both EBM and SLM built components utilizing a 150 kg load.

Several different EBM components were built. These included cylinders measuring 12 mm diameter and 68 mm length which allowed tensile specimens to be fabricated along with other analytical coupons for optical and electron microscopy. Smaller samples measuring 1 cm square and 2 cm in height were also built. Correspondingly, SLM components consisted of rectangular plates measuring 20 mm wide (and 3 mm thick) by 120 mm in length, which also facilitated the fabrication of tensile specimens. Tensile test data was obtained using an Instron 500R tensile machine and a strain rate of 3 x 10^{-3}s^{-1}. Specimens were strained to failure and the fracture surfaces examined in the FE-SEM.

RESULTS AND DISCUSSION

Figure 3(a) illustrates the microindentation hardness measurements made on the starting Ti-6Al-4V powders. For the EBM powder illustrated in Figures 2 and 3, the average Vickers

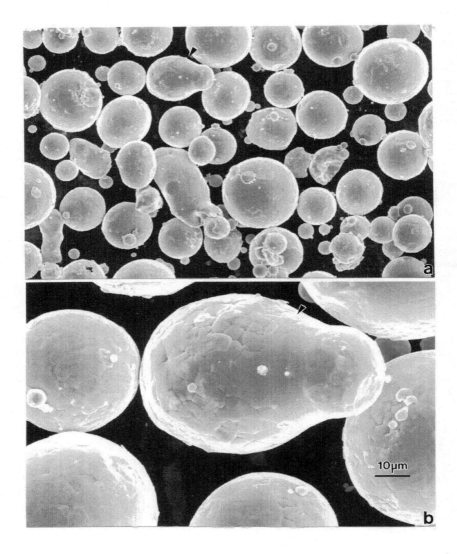

Figure 2. SEM view of the Ti-6Al-4V EBM system powder (a). (b) shows a magnified view of the sintered particles shown at arrow in (a).

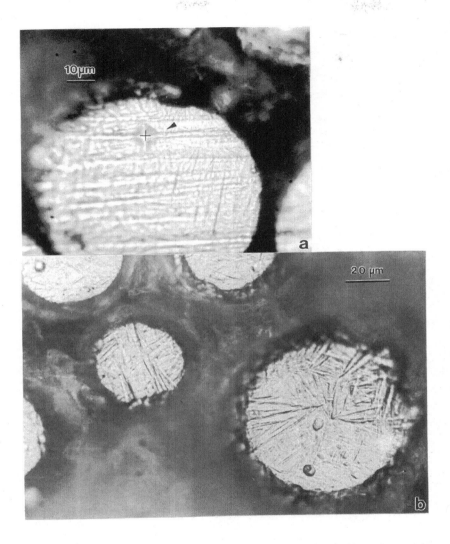

Figure 3. Optical micrographs of embedded, polished and etched Ti-6Al-4V powder particles. (a) Polished particle with Vickers microindentation (arrow). (b) Etched particles showing acicular α microstructure.

microindentation hardness was HV 52 while for the smaller SLM Ti-6Al-4V powder the average microindentation hardness was found to be HV 42. Correspondingly, the Vickers microindentation hardness for cylindrical built samples represented in the section view of Figure 4(a) varied from HV 46 near the bottom of the 68 mm build to HV 36 near the top, corresponding to a somewhat finer α acicular, basketweave-type or Widmanstätten/platelet microstructure at the bottom. These features are consistent with LENS-built specimens described by Wu, et al.[7] Figure 4(b) illustrates a so-called lack-of-fusion defect region in a small (2 cm^3) build which was polished and etched to reveal the α acicular microstructure in the FESEM.

It might be noted that while the Vickers microindentation hardnesses for the EBM builds varied from HV 36 to HV 50, the corresponding Rockwell C-scale hardnesses varied from HRC 37 to HRC 55. This compares with conventional wrought Ti-6Al-4V products. Hardnesses for the SLM builds tended to be in the higher hardness range: HRC 48 to HRC 53. Figure 5 shows for comparison build sample side views of the typical microstructures observed by optical microscopy along with corresponding views observed by transmission electron microscopy. Figure 5(a) shows the typical EBM α acicular basketweave microstructure while Figure 5(b) shows the typical SLM α' martensite microstructure which results by an order of magnitude faster beam scan rate which contributes more rapid cooling to produce the hcp α' martensite. The TEM views in Figures 5(c) and (d) illustrate the finer microstructure details. Especially notable in Figure 5(c) is the very high dislocation density (~ 10^{10} cm^{-2}) which results by the intrinsic cooling.

Figure 6 shows for comparison wrought Ti-6Al-4V α acicular microstructure having an average hardness of HRC 48 (Figure 6(a)) compared with an EBM Ti-6Al-4V α acicular microstructure having an average hardness of HRC 54 (Figure 6(b)) where the α platelet thickness is slightly smaller than the wrought (~2 μm versus ~3 μm), and somewhat more regular (more regular basketweave).

Figure 7 shows a comparative summary of the mechanical behavior for two specially fabricated wrought Ti-6Al-4V billets, a cast and HIPed Ti-6Al-4V sample from data in Niinomi[10], and the experimental EBM and SLM RP samples. The circled numbers represent average HRC values. The horizontal arrow to the right of the equiaxed α/β wrought billet represents the average yield stress and corresponding Rockwell C-scale hardness for commercial, wrought Ti-6Al-4V. Figure 6(a) is representative of the α-wrought Ti-6Al-4V microstructure while Figure 6(b) shows the α-acicular basketweave microstructure for the EBM built samples represented by an average hardness of HRC 53. The comparatively different elongations for the EBM versus the SLM builds represented in Figure 7 (~20% versus 4% respectively) are characterized by significant fracture surface feature differences as shown in Figure 8. These differences are also illustrated in the comparative α and α' martensite platelet microstructures shown in Figure 5(a) and (b). These variations in mechanical behavior and associated variations in microstructure through build parameter variations have also been illustrated for LENS by Wu, et al.[7]

While Figure 7 illustrates the similarity of EBM and α-wrought product mechanical behavior Figure 6 confirms the corresponding microstructural similarities. The EBM process can produce defects ranging from spherical pores (~10-30 μm) diameter to larger regions of lack-of-fusion interlayer porosity as illustrated in the extreme by Figure 4(b). However in

Figure 4. (a) Optical metallographic view showing acicular, α-plate (Widmanstätten) microstructures ~ 1 cm from the top of a 68 mm length build. (b) FESEM view of polished and etched Ti-6Al-4V particle in a lack-of-fusion defect zone showing the α plate microstructure.

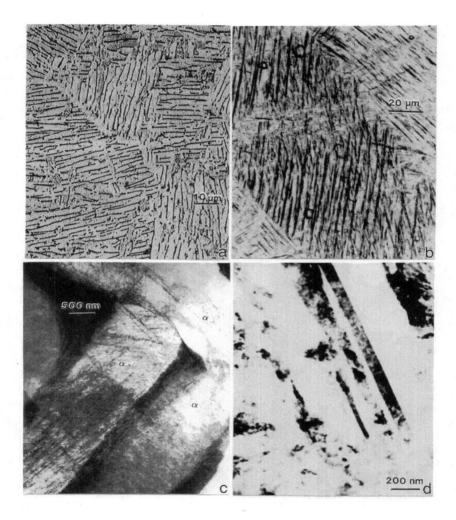

Figure 5. Comparison of EBM and SLM build microstructures. (a) Optical metallographic view of EBM build side section. (b) Optical metallographic view of SLM section. (c) TEM image for EBM section. Dark regions separating α grains are β. (d) TEM image for SLM section in (b) (After Murr, et al.[9] by permission from Elsevier Ltd.).

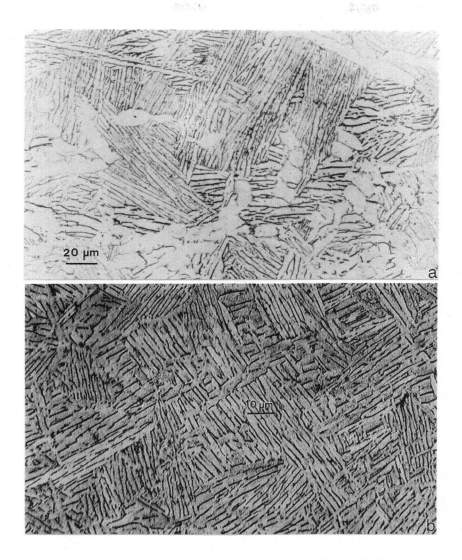

Figure 6. Optical metallographic views of a commercial, wrought Ti-6Al-4V alloy microstructure (a), and an EBM-produced Ti-6Al-4V alloy microstructure (b).

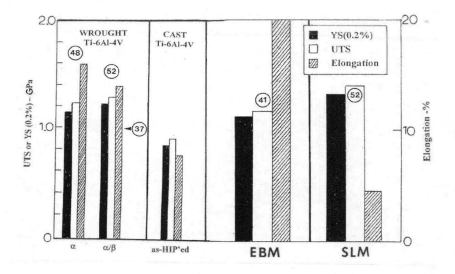

Figure 7. Comparative mechanical properties for fabricated Ti-6Al-4V alloy components: ultimate tensile strength (UTS), 0.2% engineering offset yield stress (YS) and elongation (%). Average Rockwell C-scale hardness (HRC) is shown circled. (After Murr, et al.[9] by permission from Elsevier Ltd.)

applications such as those involving metal fatigue where even small defects can act as crack nuclei, HIPing at ~900°C for 2h at 0.1 GPa pressure can effectively heal these porosities as is routinely accomplished for both wrought and cast Ti-6Al-4V alloy products.

Despite concerns for build-related porosities in EBM and SLM layer-manufactured components, the range of microstructure and mechanical properties achievable through variations of build parameters (such as beam power and scan speed) can provide for the manufacture not only of complex components with graded compositions and properties, but also building of complex, open geometries having graded densities can be achieved as well. So-called engineered porosities can also be produced as recently describe by Krishna, et al.[8] These prospects pose unprecedented, direct manufacturing capabilities not only for Ti-6Al-4V alloy systems but virtually any metal or alloy system where appropriate precursor powders are available.

CONCLUSIONS

Both EBM and SLM appear to be viable processes for digital manufacturing of complex, custom-designed components layer-by-layer, especially for biomedical and aerospace systems. For Ti-6Al-4V alloy SFF, EBM and SLM layered manufacturing can exceed wrought processing capabilities with minimum waste and near net shape fabrication.

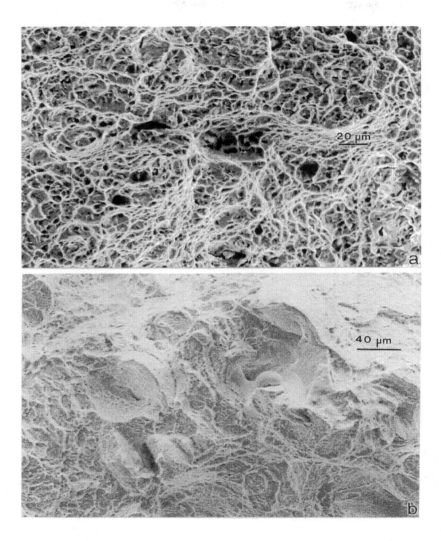

Figure 8. Field emission SEM images showing fracture surface features for EBM (a) and SLM (b). Ductile dimple features are much more prominent in (a) than in (b), which show a quasi-brittle-like fracture surface.

ACKNOWLEDGMENTS

This research was supported in part by Mr. and Mrs. MacIntosh Murchison Chair Endowments at the University of Texas at El Paso. We thank Shane Collins of EOS North America, Inc. for providing SLM fabricated specimens for analysis. We have used previously published figures or figure portions with permission from Elsevier Ltd. Acknowledgments have been made in specific figure captions.

REFERENCES

[1] J. E. Smugeresky, Laser Engineered Net Shaping Process-Optimization of Surface and Microstructures, *Metal Powder Rep.,* **53**(a), 57-68 (1998).

[2] P. A. Kobryn, E. H. Moore, and S. L. Semiatin, The Effect of Laser Power and Transverse Speed on Microstructure Porosity and Build Height in Laser-Deposited Ti-6Al-4V, *Scripta Mater.,* **43**, 299-305 (2000).

[3] C. K. Chuna, K. F. Leong, and C. S. Lim, Rapid Prototyping: Principles and Applications, 2nd Edition, World Scientific, Singapore (2003).

[4] I. Gibson (Ed.), Advanced Manufacturing Technology for Medical Applications, J. Wiley & Sons, Ltd., London (2005).

[5] J. Hiemenz, Electron Beam Melting, *Adv. Mater. Processes,* **165**, 45-46 (2007).

[6] S. L. Semiatin, P. A. Kobryn, E. D. Roush, D. V. Furrer, T. E. Howson, R. R. Boyer, and D. J. Chellman, Plastic Flow and Microstructure Evolution During Thermomechanical Processing of Laser-Deposited Ti-6Al-4V Preforms, *Metall. Mater. Trans.;* **32A**, 1801-1811 (2001).

[7] X. Wu, J. Liang, J. Mei, C. Mitchell, P. S. Goodwin, and W. Voice, Microstructures of Laser-Deposited Ti-6Al-4V, *Mater. Design,* **25**, 137-144 (2004).

[8] V. Krishna, W. Xue, S. Bose, and A. Bandyspadhyay, Engineered Porous Metals for Implants, *JOM*, May, 45-48 (2008).

[9] L. E. Murr, S. A. Quinones, S. M. Gaytan, M. I. Lopez, A. Rodela, E. Y. Martinez, D. H. Hernandez, E. Martinez, F. Medina, R. B. Wicker, *J. Mech. Behavior of Biomed. Mater.*, in press (2008).

[10] M. Niinomi, Mechanical Biocompatibilities of Titanium Alloys for Biomedical Applications, *J. Mech. Behavior Biomed. Mater.*, **1**, 30-42 (2008).

MECHANICAL PROPERTIES OF IMPLANT RODS MADE OF LOW-MODULUS β-TYPE TITANIUM ALLOY, Ti-29Nb-13Ta-4.6Zr, FOR SPINAL FIXTURE

Kengo Narita[1, 2], Mitsuo Niinomi[1], Masaaki Nakai[1], Toshikazu Akahori[1],
Harumi Tsutsumi[1], Kazuya Oribe[2], Takashi Tamura[2], Shinji Kozuka[2] and Shizuma Sato[2]
[1]Institute for Materials Research, Tohoku University, Sendai 980-8577
[2]Showa Ika Kogyo Co., Ltd., Toyohashi 441-8026

ABSTRACT

Implanting a spinal fixture by using implant rods is one of the effective treatments for spinal diseases. Most of the implant rods are made of the Ti-6Al-4V ELI alloy (Ti64). However, some problems regarding using Ti64 rods have been pointed out; these rods contain vanadium, which is highly toxic to the human body, and exhibit a considerably higher Young's modulus than the cortical bone. The Ti-29Nb-13Ta-4.6Zr alloy (TNTZ) developed by the authors exhibits good biocompatibility due to its nontoxic- and nonallergic- components. Further, TNTZ has a lower Young's modulus as compared to Ti64. It is expected that implant rods made of TNTZ exhibiting low Young's modulus will prevent the accelerated degeneration of the segment adjacent to a fusion.

The mechanical properties of TNTZ rods subjected various heat treatments are comprehensively evaluated through tensile tests, Young's modulus measurement, and four-point bending fatigue tests based on the ASTM F2193 standard and compared with those of Ti64 rods.

Among the TNTZ rods used in this study, the TNTZ rod aged at 673 K ($Rod_{673 K}$) exhibits the highest tensile strength because of the precipitation of the isothermal ω-phase and α-phase in it. The tensile strength of $Rod_{673 K}$ exceeds that of Ti64, while the increase in Young's modulus of the former is suppressed below that of the latter. With regards to four-point bending fatigue strength, no advantages are obtained from every TNTZ rods in comparison with Ti64 rod.

INTRODUCTION

The spinal fixture system comprising rods and hooks made of stainless steel developed by Harrington[1] is used for the therapeutic treatment of scoliosis. This system is also used for treating other spinal diseases such as spondylolisthesis and spinal fractures[2]. Recently, a spinal fixture system comprising rods and screws (PSS) has been used for treating such as diseases[3]. The degeneration of the segment adjacent to a fusion may be accelerated by the PSS because the rigidity of the fused section is higher than that of the cortical bone[4, 5] due to the presence of PSS in the former.

Most of the implant rods used in the PSS are made of the Ti-6Al-4V ELI alloy (Ti64). However, some problems regarding Ti64 rods have been pointed out; these rods contain vanadium, which is highly toxic to the human body, and exhibit a considerably higher Young's modulus (around 110 GPa)[6] than the cortical bone (around 10–30 GPa)[7]. The Ti-29Nb-13Ta-4.6Zr alloy (TNTZ) developed by the authors exhibits good biocompatibility due to its nontoxic and nonallergic components, and Young's modulus of solutionized TNTZ is lower than that of Ti64[8, 9]. It is expected that implant rods made of TNTZ exhibiting low Young's modulus will prevent the accelerated degeneration of the segment adjacent to a fusion.

Implants rods should exhibit superior mechanical properties. If they break in vivo, it may lead to serious complications in the patients. Therefore, implant rods are subjected to various heat treatments for improving their mechanical properties. In this study, the mechanical properties of TNTZ rods subjected to various heat treatments are comprehensively evaluated through tensile tests, Young's modulus measurement, and four-point bending fatigue tests based on the ASTM F2193 standard[10] and compared with those of a Ti64 rod.

EXPERIMENTAL PROCEDURE

Material Preparation

The material used in this study is a hot-rolled coil of TNTZ with a diameter of 8.2 mm. This coil is cold worked to form bars with a diameter of 7.5 mm and a length of 300 mm. The bars are solutionized at 1063 K for 3.6 ks in air, followed by water quenching. Some of these solutionized bars are then aged at 673 or 723 K for 259.2 ks in air, followed by water quenching. After the heat treatments, the bars are machined to form rods with a diameter of 5.0 mm, followed by blast finishing. The TNTZ rods that are as-solutionized, aged at 673 K for 259.2 ks, and aged at 723 K for 259.2 ks are referred to as Rod_{ST}, $Rod_{673 K}$, and $Rod_{723 K}$, respectively. The chemical compositions of the TNTZ rods used in this study are listed in Table I. No significant increase in the oxygen and nitrogen contents is observed after exposure of the rods to air during the heat treatments. Equiaxial grains with grain diameters in the range of 5–15 μm are observed in the TNTZ rods by optical microscopy. A conventional α + β-type titanium alloy of Ti64, which is standardized according to ASTM F136[11], used for biomedical applications is used for comparison.

Table I. Chemical compositions of TNTZ rods subjected to various heat treatment

	Nb	Ta	Zr	O	N	Ti (mass %)
Rod_{ST}	30.5	13.0	4.81	0.0781	0.0087	Bal.
$Rod_{673 K}$	30.5	13.0	4.82	0.0732	0.0074	Bal.
$Rod_{723 K}$	30.5	13.0	4.77	0.0755	0.0076	Bal.

Observation of Microstructure

The constitutional phases of the TNTZ rods subjected to various heat treatments were identified by X-ray diffraction (XRD). The specimens for XRD were obtained by cutting out disks with a diameter of 5 mm and a thickness of 2 mm from the rods. Then, the disks were wet polished using waterproof emery papers up to #1500. The XRD analysis was carried out using a Cu-Kα radiation source at a voltage of 40 kV and a current of 150 mA. The microstructures of the specimens were observed by transmission electron microscopy (TEM). Thin discs with a diameter of 2.9 mm and a thickness of 0.5 mm were prepared for TEM. Then, they were wet polished using waterproof emery papers up to #1500. After the polishing, their thicknesses were in the range of 0.05–0.1 mm. Then, they were ion milled for further thinning. The TEM analysis was carried out at an accelerating voltage of 200 kV.

Mechanical Tests

The specimens for the tensile tests were cut from the rods, machined to form specimens as shown in Figure 1, and wet polished using waterproof emery papers up to #1500. The tensile tests were carried out using an Instron-type machine with a crosshead speed of 8.33×10^{-6} m s^{-1} in air at room temperature. In this test, the strain was measured by using a strain gage attached to the specimens.

The specimens for Young's modulus measurement were cut from the rods to form short rods with a diameter of 5 mm and a thickness of 50 mm, followed by wet polishing using waterproof emery papers up to #1500. Young's moduli of the specimens were measured by the free resonance method.

In order to evaluate the suitability of TNTZ as a material for implant rods, the four-point bending fatigue tests based on the ASTM F2193 standard were carried out. The specimens for the four-point bending fatigue tests were cut from the rods to form short rods with a diameter of 5 mm and a length of 100 mm, followed by blast finishing. The outer and inner span lengths of the specimens were set at 76 and 25 mm, respectively. Then, the tests were carried out at a frequency of 8 Hz with a stress ratio of R = 0.1 in air at room temperature.

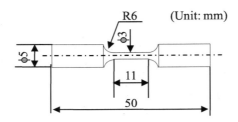

Figure 1. Geometry of specimen for tensile tests

EXPERIMENTAL RESULTS AND DISCUSSION

Microstructures of TNTZ Rods

The XRD profiles of the TNTZ rods subjected to various heat treatments are shown in Figure 2. The peaks of only the β-phase are detected in Rod_{ST}. On the other hand, the peaks of the α-phase are detected in addition to those of the β-phase in $Rod_{673\,K}$ and $Rod_{723\,K}$. Figure 3 shows the bright-field (BF) images, the selected-area electron diffraction (SAD) patterns, and the key diagrams (KD) of the TEM observation of the TNTZ rods subjected to various heat treatments. No precipitated phases are identified in the BF image of Rod_{ST}. The diffraction pattern of only the β-phase is obtained in the SAD pattern of Rod_{ST}. In contrast to Rod_{ST}, precipitated phases are identified in the BF images of both $Rod_{673\,K}$ and $Rod_{723\,K}$. In addition to the diffraction pattern of the β-phase, the diffraction patterns of the ω- and α-phases are also obtained, as shown in the SAD pattern of $Rod_{673\,K}$. This result indicates that the volume fraction of the ω-phase is too small to be detected by XRD. The diffraction pattern of only the α-phase is obtained as a precipitated phase in the SAD pattern of $Rod_{723\,K}$. The area of this precipitated phase is slightly larger than that in the BF image of $Rod_{673\,K}$.

Thus, Rod_{ST} consists of a single β-phase. In addition to the β-phase, the isothermal ω- and α-phases are identified in $Rod_{673\,K}$. On the other hand, only the α-phase is identified to be precipitated in $Rod_{723\,K}$. These results show a tendency similar to that observed in the results of our previous study[12].

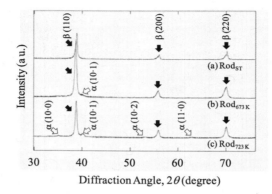

Figure 2. X-ray diffraction profiles of (a) Rod$_{ST}$, (b) Rod$_{673\,K}$, and (c) Rod$_{723\,K}$

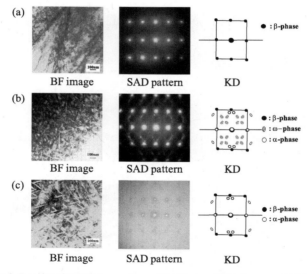

Figure 3. Results of TEM observations of (a) Rod$_{ST}$, (b) Rod$_{673\,K}$, and (c) Rod$_{723\,K}$

Mechanical Properties of TNTZ Rods

Figure 4 shows the comparison of the tensile properties of each TNTZ rod and the Ti64 rod; the tensile properties of the Ti64 rod have been obtained from a literature[13]. The tensile strength and 0.2% proof stress of the TNTZ rods increase in the following order: $Rod_{ST} < Rod_{723\,K} < Rod_{673\,K}$. However, the elongation of the rods decreases in the following order: $Rod_{ST} > Rod_{723\,K} > Rod_{673\,K}$. The tensile properties of the TNTZ rods correspond to their microstructure results. The tensile strength and 0.2% proof stress are higher and the elongation is lower in $Rod_{723\,K}$ as compared to Rod_{ST} because of the precipitation of the α-phase. The tensile strength and 0.2% proof stress increase and the elongation decreases significantly by the precipitation of the ω-phase in addition to that of the α-phase in $Rod_{673\,K}$ as compared to $Rod_{723\,K}$. Young's moduli of the TNTZ rods also correspond to their microstructure results, as shown in Figure 5. That is, the precipitation of the α-phase increases Young's modulus in $Rod_{723\,K}$ as compared to Rod_{ST}. The precipitation of the α- and ω-phases increases Young's modulus more significantly in $Rod_{673\,K}$ than in $Rod_{723\,K}$. Further, the tensile strength and 0.2% proof stress of $Rod_{673\,K}$ exceed those of the Ti64 rod, while the increase in Young's modulus of the former is suppressed below that of the latter. Therefore, $Rod_{673\,K}$ is considered to be a suitable candidate for implant rods. However, a major drawback of $Rod_{673\,K}$ is its reduced fatigue strength because the precipitation of ω-phase induces embrittlement in it.

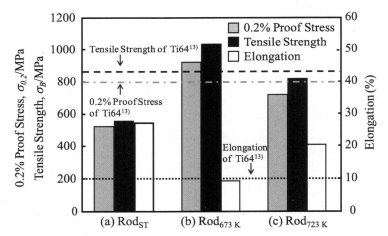

Figure 4. Comparison of tensile properties of (a) Rod_{ST}, (b) $Rod_{673\,K}$, and (c) $Rod_{723\,K}$ with those of Ti64 reported in the literature[13].

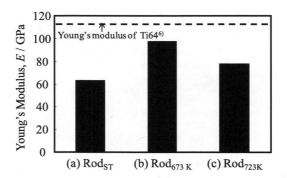

Figure 5. Comparison of Young's moduli of (a) Rod$_{ST}$, (b) Rod$_{673\ K}$, and (c) Rod$_{723\ K}$ with Young's modulus of Ti64 reported in the literature[6].

The cyclic stress-fatigue life (the number of cycles to failure) curves, that is, S-N curves, obtained from the four-point bending fatigue tests of each TNTZ rod and the Ti64 rod are shown in Figure 6. The bending fatigue strength increases in the low cycle fatigue region in the following order: Rod$_{ST}$ < Rod$_{723\ K}$ < Rod$_{673\ K}$. However, the bending fatigue strength of Rod$_{723\ K}$ is similar to that of Rod$_{673\ K}$ in this region. These tendencies also correspond to the microstructures results of the rods. The improvement in the bending strengths of Rod$_{723\ K}$ and Rod$_{673\ K}$ due to the precipitation of the ω- and α-phases and the precipitation of the α-phase, respectively, as compared to Rod$_{ST}$ may increase the crack initiation resistance and the small fatigue crack propagation resistance, leading to the improvement of their bending fatigue strengths in the low cycle fatigue region. On the other hand, the fatigue strength in the high cycle fatigue region increases in the following order: Rod$_{723\ K}$ < Rod$_{ST}$ < Rod$_{673\ K}$. However, the bending fatigue strength of Rod$_{ST}$ is similar to that of Rod$_{723\ K}$ in this region. This result indicates that the precipitation of the ω-phase affects the increase in the crack initiation and fatigue crack propagation resistances, whereas the precipitation of the α-phase does not significantly affect the increase in these resistances in the high cycle fatigue region. On the other hand, the fatigue strength of the Ti64 rod exceeds the fatigue strengths of the TNTZ rods in both low and high cycle fatigue regions.

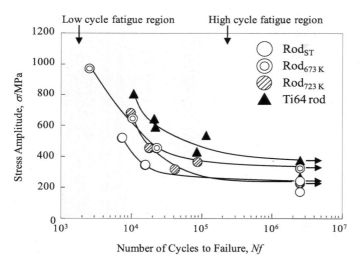

Figure 6. S-N curves of TNTZ rods subjected to various heat treatments and Ti64 rod obtained from four-point bending fatigue tests.

CONCLUSION

In this study, the mechanical properties of TNTZ rods required for implant rods have been evaluated. The following results have been obtained.

1. Rod_{ST} consists of a single β-phase. $Rod_{673 K}$ consists of the isothermal ω- and α-phases in addition to the β-phase. $Rod_{723 K}$ consists of the α-phase in addition to the β-phase.

2. Among the TNTZ rods used in this study, $Rod_{673 K}$ exhibits the highest tensile strength because of the precipitation of the isothermal ω- and α-phases in it. Further, the tensile strength of $Rod_{673 K}$ exceeds that of the Ti64 rod, while the increase in Young's modulus of the former is suppressed below that of the latter.

3. With regard to the four-point bending fatigue strength at 2.5 million cycles, no advantage is obtained from every TNTZ rods in comparison with Ti64 rod.

REFERENCES

1)   P.R. Harrington, *J. Bone Joint Surg.*, **44-A**, 591-610 (1962)
2)   P.R. Harrington, *S. Afr. J. Surg.*, **5**, 7-12 (1962)
3)   S. Nakahara, 41st Japan Medical Society of Spinal Cord Lesion, p 76 (2006)
4)   B.W. Cunningham, Y. Kotani, P.S. McNulty, A. Cappuccino, and P.C. Mcfee, *Spine*, **22**, 2655-2663 (1997)
5)   C.K. Lee, *Spine*, **13**, 375-377 (1988)
6)   G.E. Wnek and G.L. Bowlin, *Encyclopedia of Biomaterials and Biomedical Engineering Volume 1*, Marcel Dekker, Inc., p 809 (2004)
7)   R. J. Young, Y.T. Ting, and M.P. George, *Biomater.*, **18**, 1325-1330 (1997)

8)   D. Kuroda, M. Niinomi, M. Morinaga, Y. Kato, and T. Yashiro, *Mater. Sci. Eng.*, **A243**, 244-249 (1998)

9)   M. Ninomi, *Biomater.*, **24**, 2673-2683 (2003)

10) ASTM designation F2193-02: ASTM International (2002)

11) ASTM designation F136-02: ASTM International (2002)

12) T. Akahori, M. Niinomi, K. Ishimizu, H. Fukui, and A. Auzuki, J. Japan Inst. Metals, **67**, 652-660 (2003)

13) D.M. Brunette, P. Tengvall, M. Textor, and P. Thomsen, *Titanium in Medicine*, Springer-Verlag, p 35 (2001)

FUNCTIONALITY OF POROUS TITANIUM IMPROVED BY BIOPOLYMER FILLING

Mitsuo Niinomi[1], Masaaki Nakai[1], Toshikazu Akahori[1], Harumi Tsutsumi[1], Hideaki Yamanoi[1], Shinichi Itsuno[2], Naoki Haraguchi[2], Yoshinori Itoh[3], Tadashi Ogasawara[4], Takashi Onishi[4], Taku Shindoh[5]

[1]Institute for Materials Research, Tohoku University, Sendai, Japan.
niinomi@imr.tohoku.ac.jp, nakai@imr.tohoku.ac.jp, akahori@imr.tohoku.ac.jp,
Tsutsumi@imr.tohoku.ac.jp, h-yamanoi@imr.tohoku.ac.jp

[2]Department of Materials Science, Toyohashi University of Technology, Toyohashi, Japan
itsuno@tutms.tut.ac.jp, haraguchi@tutms.tut.ac.jp

[3]Hamamatsu Industrial Research Institute of Shizuoka Prefecture, Hamamatsu, Japan
itohy@iri.pref.shizuoka.jp

[4]Osaka Titanium Technologies Co., Ltd., Amagasaki, Japan
togasawara@osaka-ti.co.jp, tonishi@osaka-ti.co.jp

[5]Environment Conservation Research Institute, Tohoku University, Sendai, Japan
shindoh@env.tohoku.ac.jp

ABSTRACT

Metallic biomaterials with a higher Young's modulus than that of a bone adversely affect bone healing and remodeling. Therefore, it is important to develop the metallic biomaterials having a low Young's modulus. Porous materials are advantageous from this viewpoint because the Young's modulus decreases with increasing porosity. However, with an increase in porosity, the other mechanical properties start deteriorating simultaneously. In comparison with metallic biomaterials, polymers exhibit lower Young's moduli; therefore, a polymer filling is a possible option to improve the mechanical properties of porous materials by preventing the stress concentration at the pores without increasing the Young's modulus. Furthermore, certain polymers exhibit intrinsic biofunctionalities. Thus, a polymer filling is expected to prevent an increase in the Young's modulus and impart biofunctionalities to porous materials without deteriorating their mechanical properties. In this study, porous pure titanium (pTi) with a porosity of 22%–50% and filled with a medical polymer (polymethylmethacrylate: PMMA) (pTi/PMMMA) was firstly fabricated. The effects of the PMMA filling on the tensile strength of pTi were then investigated. However, the deterioration of mechanical properties was not satisfactorily prevented because of the poor interfacial adhesiveness between the titanium particles and the medical polymer. Therefore, in the present study, silane-coupling treatment was employed in order to improve the interfacial adhesiveness, and silane-coupling-treated (Si-treated) pTi filled with PMMA (Si-treated pTi/PMMA) was fabricated. The effect of the silane-coupling treatment on the mechanical properties of pTi/PMMA was investigated.

It is found that the PMMA filling improves the tensile strength of pTi that has a porosity of over 40%. The tensile strengths of Si-treated pTi/PMMA are greater than those of pTi and pTi/PMMA. In the fractographs of pTi/PMMA obtained after the tensile test, the detachment of titanium particles from PMMA is observed; this occurs because of poor interfacial adhesiveness between titanium particles and PMMA. However, in the case of Si-treated pTi/PMMA, the interfacial adhesiveness between titanium particles and PMMA is improved by the silane-coupling treatment. This leads to the dispersion of the stress concentration at the necks between particles, resulting in an improvement in the tensile strength of pTi. On the other hand, the PMMA filling hardly affects the Young's modulus of pTi because the Young's modulus of PMMA is lower than that of pTi.

INTRODUCTION

New biocompatible β-type titanium alloys with a low Young's modulus have been recently developed for biomedical applications [1, 2, 3], because a low Young's modulus comparable to that of the human bone is said to be advantageous for healthy bone, hearing, and remodeling applications [4, 5]. As a result, titanium alloys with Young's moduli in the range of 40–60 GPa [6, 7] have been developed, although these values are still higher than the Young's modulus of the human bone (10–30 GPa) [8, 9].

One of the approaches to develop materials having a lower Young's modulus is to use porous materials. For example, porous titanium and its alloys, in comparison with the bulk alloys, have been reported to exhibit lower Young's moduli [10, 11]. However, their mechanical properties deteriorate drastically with an increase in porosity [10, 11]. This deterioration may be attributed to the stress concentration at the pores. Thus, the inhibition of the stress concentration by filling the pores with certain materials is likely to improve the mechanical properties of porous materials. In such a case, materials with a low Young's modulus should be selected for filling in order to prevent an increase in the Young's modulus. Young's modulus and strength are mechanical functionalities and may be collectively referred to as mechanical biocompatibilities in the broad sense. Furthermore, depending on the type of the filled materials, biofunctionalities, which are not intrinsically present in metallic biomaterials, can be imparted to the porous materials in addition to improving their mechanical properties, which are mechanical functionalites. Further, tight bonding is generally desired between porous material and the filling used. Therefore, the interface between them is subjected to some treatments.

In this study, porous pure titanium (pTi) and a medical polymer (polymethylmethacrylate: PMMA) were selected as porous material and filling material, respectively. Then, pTi filled with PMMA (pTi/PMMA) was fabricated. Furthermore, silane-coupling treatment was employed in order to improve the interfacial adhesiveness, and silane-coupling-treated pTi filled with PMMA (Si-treated pTi/PMMA) was fabricated. The effects of the PMMA filling on the mechanical properties of pTi and the effects of the silane-coupling treatment on the mechanical properties of pTi/PMMA were investigated. The evaluations of their mechanical properties in vitro and in vivo should be done, but they are future works.

Table 1. Particle diameter range, porosity, and sintering conditions(sintering temp., time, and pressure) of pTi.

Material	Particle diameter range (μm)	Porosity (%)	Sintering temp. (K)	Sintering time (ks)	Sintering pressure (Pa)
pTi45-22	< 45	22	1423	3.6	100
pTi45-35	< 45	35	1173	3.6	100
pTi150-27	45 – 150	27	1573	36	0.01
pTi150-38	45 – 150	38	1623	3.6	100
pTi150-45	45 – 150	45	1323	3.6	100
pTi250-45	150 – 250	45	1623	3.6	100
pTi250-50	150 – 250	50	1423	3.6	100

EXPERIMENTAL PROCEDURES

Gas-atomized pure titanium powders (Ti: bal., O: ≤ 0.13 mass%, Sumitomo Titanium Co., Japan) were used as the starting material. The powders were sieved into three different particle sizes with diameters in the range of 0–45 μm, 45–150 μm, and 150–250 μm and were filled in a disk-shaped zirconia tap with a diameter of 100 mm and a thickness of 2 mm. Subsequently, the powders were sintered at a constant temperature of 1173–1623 K under a reduced pressure of 1×10^2 Pa for 3.6 ks or 1×10^{-2} Pa for 36 ks, followed by furnace cooling. Consequently, seven types of porous sintered disk-shaped specimens were obtained. The particle diameter range, porosity, and sintering conditions (sintering temperature, sintering time, and sintering pressure) of porous pure titanium are shown in Table 1. Then, for tensile testing, 4–6 plate specimens (pTi tensile specimens) were prepared from the disk-shaped specimens by water jet machining. These tensile specimens had a cross section of 3×2 mm^2 and a parallel part of 13 mm (gauge length: 10 mm).

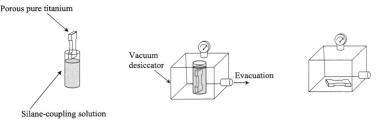

(1) A porous pure titanium is soaked in a silane coupling solution.

(2) Air babbles in the pores of the porous pure titanium are removed under a reduced pressure.

(3) The porous pure titanium is dried under a reduced pressure.

Figure 1. Schematic drawing of method of silane-coupling treatment.

Half of the pTi tensile specimens were subjected to the silane-coupling treatment before filling PMMA, and the other half of the pTi tensile specimens were filled with PMMA without silane coupling treatment. The silane-coupling treatment was carried out according to the method schematically shown in Figure 1. Firstly, the pTi tensile specimens were immersed in a glass tube containing the solution of a silane-coupling agent; the solution was composed of 98% acid solution (pH = 4.2, maintained constant) and 2% silane coupling agent of 3-Methacryloxypropyltrimethoxysilane. The glass tube was then kept in a camber under a reduced pressure of 0.05 MPa for 6 ks in order for the solution of the silane-coupling agent to penetrate all the pores. Then, after a fixed time, the tensile specimens were taken out from the glass tube and dried at 313 K for 86.4 ks under a pressure of 0.01 MPa in a vacuum chamber.

Figure 2 schematically shows the fabrication process used in this study for fabricating the pTi tensile specimens filled with PMMA (pTi/PMMA tensile specimens) and the pTi tensile specimens subjected to silane-coupling treatment (Si-treated pTi/PMMA tensile specimens). pTi or Si-treated pTi tensile specimens were soaked in an methylmetacrylate (MMA) solution mixed with 5 mol% AIBN. In order to remove the air bubbles in the pores of pTi, the MMA solution was placed in a chamber under a reduced pressure of 0.05 MPa at room temperature for 3.6 ks. These polymerization conditions were optimized in our previous work [12, 13]. Subsequently, polymerization was carried out in air by heating the MMA solution at a constant temperature of 313 K. Then, the excess PMMA was removed from the pTi/PMMA or Si-treated pTi/PMMA tensile specimens by mechanical machining. The PMMA filling rate was evaluated by an image analysis technique, and a high filling rate, larger than 98%, was obtained in this study.

The microstructure of the cross section of pTi was observed using an optical microscopy. For the microstructural observation, the specimens were mirror finished by buff polishing after being polished using a #1500 waterproof emery paper. Furthermore, X-ray diffraction (XRD) was carried out at a scanning rate of 2° min^{-1} using a Cu target with an accelerating voltage of 40 kV and a current of 150 mA.

The tensile specimens were then polished using #600 waterproof emery papers. The tensile test was carried out using an Instron-type machine with a crosshead speed of 8.33×10^{-6} m s^{-1} in air at room temperature. The fracture surface of each tensile-tested specimen was treated by carbon evaporation and then observed by SEM.

EXPERIMENTAL RESULTS AND DISCUSSION
Microstructure of pTi

The optical microstructure obtained from the cross section of each pTi is shown in Figure 3. The contact area between the particles depends on the particle size and the sintering conditions. Among the specimens with the same particle size, the contact area increases with an increase in the sintering temperature and sintering time. Also, the sintering tends to proceed faster with decreasing particle diameter. Particularly, the original shape of the particles in pTi45-22 and pTi150-27 disappears by sintering.

The XRD patterns obtained from pTi are shown in Figure 4. It is confirmed that only the peaks of the α-phase are obtained from every specimen. This result indicates that they comprise a single α-phase.

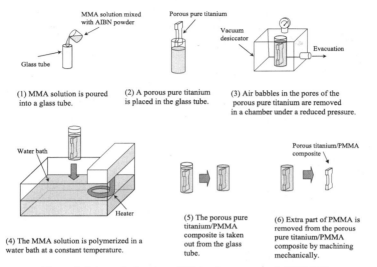

Figure 2. Schematic drawing of fabrication process of pTi/PMMA.

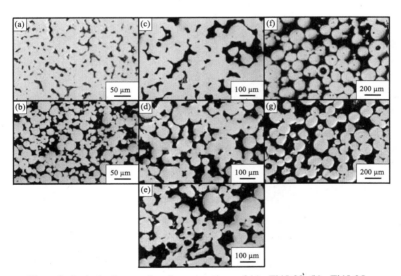

Figure 3. Optical micrographs of cross-sections of (a) pTi45-22, (b) pTi45-35, (c) pTi150-27, (d) pTi150-38, (e) pTi150-45, (f) pTi250-45, and (g) pTi250-50.

Tensile strength and Young's modulus of pTi, pTi/PMMA, and Si-treated pTi/PMMA

Figure 5 shows the tensile strengths of pTi and pTi/PMMA ($\sigma_{B(pTi/PMMA)}$ and $\sigma_{B(pTi)}$) and the tensile strength ratio ($\sigma_{B(pTi/PMMA)}/\sigma_{B(pTi)}$) plotted against their porosities. The tensile strengths of both pTi and pTi/PMMA decrease with increasing porosity. This result implies that the tensile strength of pTi and pTi/PMMA can be controlled by varying the porosity of pTi. When the porosity is in the range of approximately 30%–40%, the tensile strength of pTi is comparable to that of the human bone (60–150 MPa [14, 15]). Furthermore, although the effect of the PMMA filling on the tensile strength of pTi is unclear from the relationship between porosity and $\sigma_{B(pTi/PMMA)}$ or $\sigma_{B(pTi)}$, it is observed that the tensile strengths of pTi/PMMA are higher than those of pTi with high porosity. On the other hand, the tensile strengths of pTi/PMMA are comparable to those of pTi with low porosity. In the relationship between porosity and $\sigma_{B(pTi/PMMA)}/\sigma_{B(pTi)}$, the value above 1.0 implies that there is an improvement in the tensile strength of pTi due to the PMMA filling present in the pores. It is clearly observed that the tensile-strength ratio increases with an increase in porosity by over 40% approximately, while it is almost constant at 1.0 for a porosity of up to around 40%. This result indicates that the effect of the PMMA filling on the improvement in the tensile strength of pTi is likely to appear in the low-tensile-strength range of pTi. The upper limit of this range corresponds to the tensile strength of PMMA. In other words, the tensile strength of pTi/PMMA is dominated by that of PMMA itself in such a low-tensile-strength range of pTi.

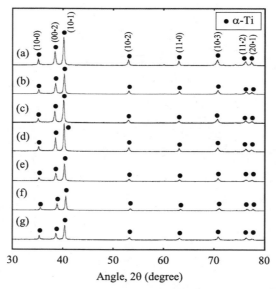

Figure 4. XRD patterns obtained from (a) pTi45-22, (b) pTi45-35, (c) pTi150-27, (d) pTi150-38, (e) pTi150-45, (f) pTi250-45, and (g) pTi250-50.

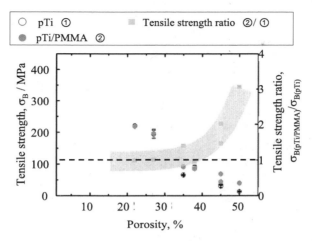

Figure 5. Tensile strengths of pTi and pTi/PMMA as a function of porosity.

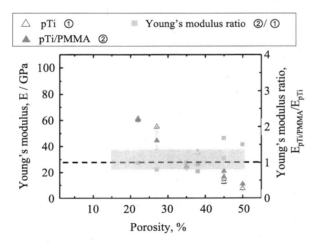

Figure 6. Young's moduli of pTi and pTi/PMMA as a function of porosity.

Figure 6 shows the Young's moduli of pTi and pTi/PMMA (E_{pTi} and $E_{pTi/PMMA}$), and the ratio of Young's moduli ($E_{pTi/PMMA}/E_{pTi}$) plotted against their porosities. The Young's modulus of pTi decreases with increasing porosity. Moreover, the Young's modulus of pTi/PMMA exhibits a trend similar to that of pTi. This trend can be clearly understood from the relationship between porosity and $E_{pTi/PMMA}$ /E_{pTi}. This result indicates that the Young's modulus of pTi/PMMA may mainly depend on that of pTi itself because the Young's modulus of PMMA is quite low. According to literature [8, 9], the Young's modulus of the human bone is reported to be approximately 10–30 GPa. The Young's modulus of pTi/PMMA fabricated in this study is also approximately 10–60 GPa. This result indicates that by controlling the porosity, it may be possible to approximate the Young's modulus of porous pure titanium to that of the human bone.

Figures 7 and 8 show the tensile strengths of pTi/PMMA and Si-treated pTi/PMMA, and the Young's moduli of pTi/PMMA and Si-treated pTi/PMMA. As a reference, the tensile strength (50–80 MPa) [16] and Young's modulus (2–4 GPa) [16, 17] ranges of PMMA are also shown in these figures. The tensile strength of each Si-treated pTi/PMMA is greater than that of each pTi/PMMA. For Si-treated pTi/PMMA, the increase in the tensile strength caused by the PMMA filling can be seen in pTi showing the relatively greater tensile strength.

The Young's modulus of Si-treated pTi/PMMA is nearly equal to that of pTi/PMMA with relatively lower porosity, but is relatively greater than that of pTi/PMMA with relatively higher porosity.

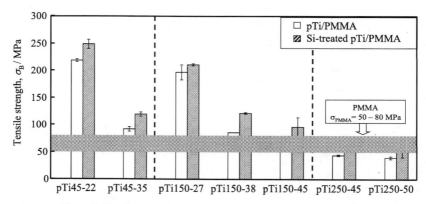

Figure 7. Tensile strengths of pTi/PMMA and Si-treated pTi/PMMA.

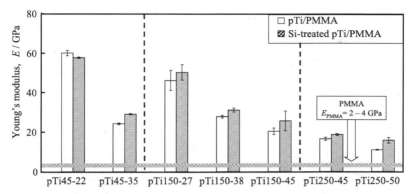

Figure 8. Young's moduli of pTi/PMMA and Si-treated pTi/PMMA.

Fracture surface of pTi, pTi/PMMA, and Si-treated pTi/PMMA

Figure 9 shows the SEM fractographs of representative pTi (pTi250-45), pTi/PMMA (pTi250-45/PMMA), and Si-treated pTi/PMMA (Si-treated pTi250-45/PMMA). The trace of the detachment of a Ti particle (spheroidal cavity) can be observed in the part of PMMA on the fracture surface of pTi/PMMA. This result suggests that the chemical bonding between a titanium particle and PMMA is not very strong. Therefore, it is difficult to release the stress concentrated at the pores of pTi/PMMA fabricated in this study. However, the fracture of PMMA can be observed on the fracture surface of Si-treated pTi/PMMA. This fact suggests that an improvement of the adhesion of titanium particles with PMMA is achieved, and the stress concentration at the pores of Si-treated pTi-PMMA is released leading to a further increase in the tensile strength of pTi/PMMA.

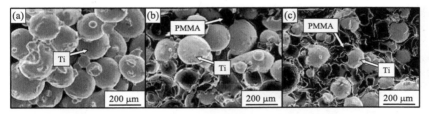

Figure 9. SEM fractographs of (a) pTi250-45, (b) pTi250-45/PMMA,
and (c) Si-treated pTi250-45/PMMA.

CONCLUSIONS

In this study, porous pure titanium (pTi) filled with a medical polymer (polymethylmethacrylate; PMMA) (pTi/PMMA) was fabricated. Then, the effect of the PMMA filling present in the pores on the tensile strength of pTi/PMMA was investigated. Silane-coupling treatment was employed in order to improve the interfacial adhesiveness, and silane-coupling-treated pTi filled with PMMA (Si-treated pTi/PMMA) was fabricated. Subsequently, the effect of the silane-coupling treatment on the mechanical properties of pTi/PMMA was investigated. The following results were obtained.

(1) The tensile strength of pTi/PMMA is improved by the PMMA filling only in the low-tensile-strength range of pTi. In this range, the tensile strength of pTi/PMMA is dominated by that of PMMA itself.

(2) The tensile strengths of Si-treated pTi/PMMA are greater than those of pTi and pTi /PMMA.

(3) In the fractographs of pTi/PMMA obtained after the tensile test, the detachment of titanium particles from PMMA is observed; this occurs because of poor interfacial adhesiveness between titanium particles and PMMA.

(4) In the case of the Si-treated pTi/PMMA, the interfacial adhesiveness between titanium particles and PMMA is improved by the silane-coupling treatment.

(5) The PMMA filling hardly affects the Young's modulus of pTi because the Young's modulus of PMMA is lower than that of pTi.

ACKNOWLEDGMENT

This work was supported in part by the Research Foundation for Materials Research, the Light Metal Educational Foundation, Osaka, Japan, the Global COE Materials Integration Program, Tohoku University, Sendai, Japan, the Japan Society for the Promotion of Science (JSPS), Tokyo, Japan, the Exploratory Research Program for Young Scientists (ERYS), Tohoku University, Sendai, Japan, and the inter-university cooperative research program of the Institute for Materials Research, Tohoku University, Sendai, Japan.

REFERENCES

[1]M. Niinomi, T. Hanawa, and T. Narushima, Japanese Research and Development on Metallic Biomedical, Dental, and Healthcare Materials, *JOM*, **57**, 18–24 (2005).

[2]M. Niinomi, Recent Metallic Materials for Biomedical Applications, *Metall. Mater. Trans. A*, **33A**, 477–86 (2002).

[3]H. J. Rack and J. I. Qazi, Titanium Alloys for Biomedical Applications, *Mater. Sci. Eng. C*, **26**, 1269–77 (2006).

[4]B. J. Moyen, P. J. Lahey, E. H. Weinberg, and W. H. Harris, Effects on Intact Femora of Dogs of the Application and Removal of Metal Plates, *J. Bone Joint Surg. (Am.)*, **60A**, 940–7 (1978).

[5]H. K. Uhthoff and M. Finnegan, The Effects of Metal Plates on Post-traumatic Remodeling and Bone Mass, *J. Bone Joint Surg. (Br.)*, **65B**, 66–71 (1983).

[6]D. Kuroda, M. Niinomi, M. Morinaga, Y. Kato, and T. Yashiro, Design and Mechanical Properties of New β Type Titanium Alloys for Implant Materials, *Mater. Sci. Eng. A*, **243**, 244–9 (1998).

[7]H. Matsumoto, S. Watanabe, and S. Hanada, Beta TiNbSn Alloys with Low Young's Modulus and High Strength, *Mater. Trans.*, **46**, 1070–8 (2005).

[8]J. Y. Rho, T. Y. Tsui, and G. M. Pharr, Elastic Properties of Human Cortical and Trabecular Lamellar Bone Measured by Nanoindentation, *Biomater.*, **18**, 1325–30 (1997).

[9]P. Zioupos and J. D. Currey, Changes in the Stiffness, Strength, and Toughness of Human Cortical Bone with Age, *Bone*, **22**, 57–66 (1998).

[10]I. H. Oh, N. Nomura, N. Masahashi, and S. Hanada, Mechanical Properties of Porous Titanium

Compacts Prepared by Powder Sintering, *Scr. Mater.*, 49: 1197–202 (2003).

[11]N. Nomura, T. Kohama, I. H. Oh, S. Hanada, A. Chiba, M. Kanehira, and K. Sasaki. Mechanical Properties of Porous Ti–15Mo–5Zr–3Al Compacts Prepared by Powder Sintering, *Mater. Sci. Eng. C*, **25**, 330–5 (2005)

[12]M. Nakai, M. Niinomi, T. Akahori, Y. Shinozaki, H. Toda, S. Itsuno, N. Haraguchi, Y. Itoh, T. Ogasawara, and T. Onishi, Mechanical Properties of Porous Titanium Filled with Polymethylmethacrylate for Biomedical Applications, Proc. Int. Conf. on Advanced Technology in Experimental Mechanics 2007 (ATEM'07), September 12–14, 2007, CD-ROM, OS07-1-4.

[13]M. Nakai, M. Niinomi, T. Akahori, Y. Shinozaki, H. Toda, S. Itsuno, N. Haraguchi, Y. Itoh, T. Ogasawara, and T. Onishi, Development of Fabrication Process of Porous Titanium and Polymethyl -methacrylate (PMMA) Composite Biomaterial, Proc. the 11th World Conference on Titanium (Ti-2007), June 3–7, 2007, 1497–500.

[14]J. C. Wall, S. K. Chatterji, and J. W. Jeffery, Age-Related Changes in the Density and Tensile Strength of Human Femoral Cortical Bone, *Calcif. Tiss. Intl.*, **27**, 105–8 (1979).

[15]W. R. Moore, S. E. Graves, and G. I. Bain, Synthetic Bone Graft Substitutes, *ANZ J. Surg.*, **71**, 354–61 (2001).

[16]K. Honda, Akuriru Jushi, The Nikkan Kogyo Shimbun., Ltd., Tokyo, 68 (1961).

[17]J. R. Young and P. R. Beaumont, Time-dependent failure of poly (methyl methacrylate), *Polymer*, **17**, 717–22 (1976).

Scaffolds for Tissue Engineering

FRACTURE FORCES IN FEMURS IMPLANTED WITH PMMA

Dan Dragomir-Daescu[1]*, Hilary E. Brown[1], Nadia Anguiano-Wehde[1], Sean McEligot[1], Michael J. Burke[1], Kevin E. Bennet[1], James T. Bronk[2], and Mark E. Bolander[2]

1. Division of Engineering, Mayo Clinic, Rochester, MN, USA
2. Department of Orthopedic Surgery, Mayo Medical School, Rochester, MN, USA

* Corresponding Author. Tel.: (507) 538-4946
Email: DragomirDaescu.Dan@mayo.edu (Dan Dragomir-Daescu)

Keywords: Bone Cement, Femoral Augmentation, Hip Fracture, Osteoporosis

ABSTRACT

With increasing numbers of hip fractures each year in the US, it is important to develop preventive treatment options including prophylactic reinforcements and tools to assess treatment efficacy. Thirteen cadaver femurs including five normal, two osteopenic and six osteoporotic pairs, were selected from a larger sample and were tested to fracture in a materials testing machine. Three osteoporotic femurs were injected with PMMA in the neck region while the contralateral femurs were used as controls. Photographs from two orthogonal directions captured the geometry of the femur and system prior to testing. Fracture forces and moments were measured using three load cells. The load cells were attached to a custom fixture that was designed to characterize all forces and moments during testing. A total of eight force and moment signals, along with machine head displacement and time were recorded simultaneously. A detailed biomechanical analysis of the forces and moments at fracture was performed using the measured loads and the geometry of bones in the test fixture. The analysis confirmed that the PMMA implanted osteoporotic femurs fractured at higher forces than their contralaterals. The reinforced femurs' fracture forces were similar to those of normal bones. Their fracture patterns more closely resembled normal femurs than osteoporotic femurs.

INTRODUCTION

As the U.S. population ages and life expectancy continues to rise, the number of hip fractures resulting from mild to moderate trauma (i.e., osteoporotic fractures) is projected to increase 2- to 3-fold by 2040. The lifetime risk of suffering a hip fracture is as high as 35% to 40% for Caucasian women over the age of 65 with osteoporosis [1-3]. The substantial contributions of hip fractures to excess mortality, disability, and health care costs are well demonstrated, and as an aging U.S. population pushes the number of hip fractures upward, these costs will increase dramatically [4-7]. In large clinical trials, drug and hormone therapies (with alendronate, risedronate, or parathyroid hormone) have decreased the hip fracture rate a maximum of 50% [8-13]. If we consider that (1) a 15-20% lifetime risk of hip fracture is still significant, (2) this risk may be higher in specific subpopulations, and (3) the number of hip fractures is increasing dramatically, then surgical treatments which predictably reduce the risk of hip fractures could have significant clinical value [14, 15].

Interventions that completely prevent fracture in osteoporotic bone have yet to be developed. Given the imperfect success of current medical therapies in preventing hip fractures

105

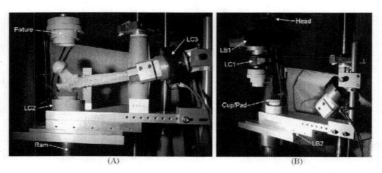

Figure 1 Experimental test apparatus and MTS material testing machine. LC1 and LC2 are the single-axis load cells; LC3 is a multiaxial load cell. (A) Proximal femur experimental set-up. (B) MTS machine.

and the significance of these injuries, we believe there is a role for surgical interventions that reduce the incidence of hip fractures. Even if limited to high-risk populations, successful surgical intervention will decrease morbidity, mortality, and socioeconomic costs. At a fundamental level, improved understanding of the forces and moments acting in the proximal femur just before fracture will form the foundation for more meaningful innovation in preventing fracture and its complications. Specifically, it will allow developing a dynamic analysis of the forces and moments acting in the femur before and during fracture. Such experimental models will be used to validate finite element models of bone strength that could provide important insights into osteoporosis diagnosis and treatment.

MATERIALS AND METHODS

Cadaveric femurs were provided by the Musculoskeletal Transplant Foundation (MTF) (Edison, NJ, USA). Proximal femurs were wrapped in saline-soaked towels and frozen until 24 hours prior to use, then thawed to room temperature. To understand the specimen's bone health, BMD was determined using a GE Lunar iDXA system (GE Healthcare, Inc., Waukesha, WI, USA). Classification of the femurs was based on the World Health Organization criteria [16]. BMD values ranged from 0.936 to 1.11 g/cm^2 for normal bones (T-score -0.4 to 1.1), 0.733 to 0.798 g/cm^2 for osteopenic bones (T-score -2.1 to -1.5), and 0.319 to 0.656 g/cm^2 for osteoporotic bones (T-score -5.5 to -2.7). A total of 47 proximal femurs were tested for strength: 10 normal, 13 osteopenic, and 24 osteoporotic (12 pairs). Thirteen of these femurs were analyzed in this work: 5 normal, 2 osteopenic, and 6 osteoporotic (3 pairs).

To fracture the femurs, a custom testing apparatus was attached to a standard mechanical testing system (MTS, Minneapolis, MN, USA) (Figs. 1A and B). The hydraulic ram moved at a velocity of 100 mm/s, assumed to represent falls resulting in hip fractures [17]. Two single-axis 22-kN load cells measured vertical forces applied to the femoral head (LC1) and greater trochanter (LC2). The distal end of the femur specimen was clamped to a third load cell (LC3). LC3, a six-channel load cell, was used to measure three force components and three moment components in the femoral shaft. LC3 can measure peak forces up to 2000 N in the axis parallel to the femoral shaft and 1000 N in the two orthogonal axes; the torque rating of load cell LC3 was 125 N·m. The greater trochanter was supported during testing by a smooth pad; the femoral head was fitted into an ultra-high molecular weight polyethylene-lined hemispherical socket inside the upper fixture of the MTS. In preliminary studies we have shown that these fixtures accurately transmit forces to the load cells and prevent bone crushing. The upper and lower fixtures were attached to the MTS test machine through linear bearings (LB1 and LB2) that permitted displacement in the horizontal plane and allowed the three load cells to characterize all forces and moments acting on the femur (Fig. 1B). Taken together, the five force components and three moment components completely described the loads imposed on the femur during strength testing at all supports. The relatively low speed of the moving parts of our test machine allowed us to neglect the inertial forces in the equilibrium equation.

We used the load cell data and geometrical measurements of the femurs in the test fixture to solve the system of static equilibrium equations to test the consistency of our measurements for each femur. The coordinate system of load cell LC3 was left-hand oriented with the z axis along the femoral shaft and x and y axes perpendicular to the shaft. LC1 and LC2 were

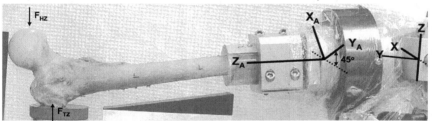

Figure 2: LC3 Coordinates (A) and machine coordinates (no subscript). The acting forces, on the trochanter and the head (F_{TZ} and F_{HZ}), are both are parallel to the machine's z-axis.

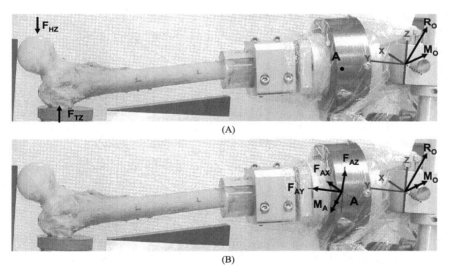

(A)

(B)

gure 3: Two separate systems for solving for M_O and R_O. In (A) the internal forces at point A are ignored. In (B), the system only includes the segment from the origin to A, thus the trochanter and head forces are neglected.

identically oriented with the z axis vertical and parallel aligned with the vertical axis of the testing machine coordinate system. To ensure that the measured forces at LC3 were accurately transformed to the machine coordinate system parallel to LC1 and LC2, we rotated the x- and y-axes by 45° and reversed the z-axis, using geometrical transformations. Then, we rotated the, now right-hand oriented, x-, y-, and z-axis to align them with the machine coordinates system. Figure 2 shows the original coordinate system of LC3 and the machine coordinate system.

Once all of the measurements were expressed in the machine coordinate system, we used two different systems of equations. One was written for the equilibrium of the femur and fixture together (Eqs. 1 and Fig. 3A) and the other one for the equilibrium of the fixture alone (Eqs. 2 and Fig. 3B). Because we measured all forces and moments, and in addition we obtained most geometry data from pictures taken before femurs were fractured we had fewer unknowns than equations. These two separate systems thus yielded several methods of solution, in addition to ways to check the validity of our assumptions, which will be discussed shortly. Using the sums of the forces and moments yielded twelve equations as follows. For the femur and fixture together:

$$
\begin{aligned}
M_{OX} &= Z_H F_{HY} - Y_H F_{HZ} + Z_T F_{TY} - Y_T F_{TZ} \\
M_{OY} &= X_H F_{HZ} - Z_H F_{HX} + X_T F_{TZ} - Z_T F_{TX} \\
M_{OZ} &= Y_H F_{HX} - X_H F_{HY} + Y_T F_{TX} - X_T F_{TY} \\
R_{OX} &= -F_{TX} - F_{HX} \\
R_{OY} &= -F_{TY} - F_{HY} \\
R_{OZ} &= -F_{TZ} - F_{HZ}
\end{aligned}
\tag{1}
$$

and for the fixture alone:

$$
\begin{aligned}
M_{OX} &= Z_A F_{AY} - Y_A F_{AZ} - M_{AX} \\
M_{OY} &= X_A F_{AZ} - Z_A F_{AX} - M_{AY} \\
M_{OZ} &= Y_A F_{AX} - X_A F_{AY} - M_{AZ} \\
R_{OX} &= -F_{AX} \\
R_{OY} &= -F_{AY} \\
R_{OZ} &= -F_{AZ}
\end{aligned}
\tag{2}
$$

where (X_A, Y_A, Z_A), (X_H, Y_H, Z_H), and (X_T, Y_T, Z_T) indicate the positions in the machine coordinate system of the center A of LC3, the center of contact H between the femoral head and the upper part of the fixture, and the center of contact T between the trochanter and the lower part of the fixture, respectively.

As mentioned, all distances were measured from two orthogonal view photographs of the system. Each photograph captured a specific femur and its alignment in the machine before fracture. All distances and angles were measured on the photographs and then scaled to real physical dimensions. The scaling was performed by comparing the distances and angles of bones with those of triangular prisms photographed next to the femurs. The scaling factors were obtained simply by dividing the measured values of the triangular blocks in the picture by its known dimensions (Fig 3.) The scaling yielded the actual distances of the system, which were especially important while calculating the moments.

Ten variables were initially considered unknown: the three components each for the reaction force R_O, and the moment at O, M_O respectively, and the force components in the x- and y-directions on the greater trochanter and femoral head. To simplify the problem, we assumed that M_{OX} was zero, because of the pin at the origin O of the coordinate system. The pin allowed the system to rotate about the x-axis. Next, we assumed that F_{HY} was zero. We also assumed that X_T and X_A were both zero, i.e. point A and the point of contact for the trochanter were on the z-y plane in machine coordinates. With these assumptions we had more equations (data) than unknowns and solved the over-determined system of equations using the least square method.

RESULTS

Tables 1 A-D show the averages, standard deviations, and ranges of the magnitudes of all forces and moments, for each femur type. The normal femurs sustained the highest fracture forces and moments. For example the sample average for F_{TZ}, the largest of the fracture forces, was 5572 N for the normal femurs, 4452 N for the osteopenic femurs and only 2122 N for the osteoporotic femurs. The PMMA reinforced osteoporotic femurs averaged $F_{TZ} = 5642$ N. This value was very similar to the sample average for normal femurs. Therefore, the PMMA augmentation procedure was effective in restoring the strength of osteoporotic bone to normal values.

After solving the system of equations, we calculated the residuals (errors) to assess how well the measured and calculated data verified the conditions for equilibrium (Eqs.1 and 2.) The residuals showed that the most accurate values were for M_{OY}, and R_{OZ}, which were obtained from the vertical forces on the greater trochanter and femoral head. Both have relative residual errors averaging 2.5 % or less. To calculate these values we used very reliable measurements for both loads and geometry. Obviously, the residual for M_{OX} was zero because M_{OX} was set to zero. This shows that the most important force and moment components in the system verify the equilibrium equations to a high degree, thus confirming the soundness of our testing procedure and measurements. For R_{OX}, and R_{OY}, we obtained much larger relative residual errors. These errors most likely originated in the assumptions that F_{TX} and F_{HX} were set to zero, while in practice they were small but non-negligible. However, the values for R_{OX}, and R_{OY}, components were also two or more orders of magnitude smaller than the applied vertical forces acting on the femur, F_{TZ} and F_{HZ}. Therefore, these components do not play an important role in the overall equilibrium of forces during fracture. In addition, for M_{OZ}, the geometry measurements used to calculate this quantity were the least reliable thus justifying another larger residual.

With simplified femur geometry, we created diagrams that showed the forces and moments for equilibrium, and the internal forces and moments at any point in the femur. Specifically, we compared the values measured by LC3, at point A and the values predicted by the diagrams. The values in the diagrams were calculated using only the directly applied forces and the test fixture and femur geometry. They were in relatively good agreement. These results will be used and compared to the results of a more detailed finite element analysis in a later study. This investigation will include the dynamic effects. Some errors in this study probably originated from the fact that the system is truly dynamic and that the femur has varying cross sectional characteristics unlike simplified beams. The finite element results of a dynamic analysis would probably be more accurate, and would eventually allow us to better quantify how well reinforcement may help in preventing hip fracture.

Table 1A: Measured and calculated external forces and moments in Normal femurs

Normal	Average	Standard Deviation	Range
F_{AX} (N)	18.79	13.62	2.93 - 35.36
F_{AY} (N)	101.75	68.68	22.36 - 186.97
F_{AZ} (N)	422.95	84.09	320.15 - 501.30
M_{AX} (N*m)	23.86	7.66	14.63 - 33.92
M_{AY} (N*m)	26.61	22.46	7.84 - 65.11
M_{AZ} (N*m)	10.72	2.77	8.38 - 15.26
F_{TX} (N)	385.13	301.20	65.18 - 719.88
F_{TY} (N)	229.10	205.75	27.85 - 512.70
F_{TZ} (N)	5571.90	970.34	4264.35 - 6656.82
F_{HX} (N)	393.54	278.86	94.89 - 710.73
F_{HY} (N)	0.00	0.00	0
F_{HZ} (N)	5183.79	822.58	4188.44 - 6314.90
R_{OX} (N)	18.79	13.62	2.93 - 35.36
R_{OY} (N)	101.75	68.68	22.36 - 186.97
R_{OZ} (N)	405.53	158.81	198.03 - 625.87
M_{OX} (N*m)	0.00	0.00	0
M_{OY} (N*m)	26.44	22.46	7.58 - 64.90
M_{OZ} (N*m)	11.31	2.47	8.56 - 14.97

Table 1B: Measured and calculated external forces and moments in Osteopenic femurs

Osteopenic	Average	Standard Deviation	Range
F_{AX} (N)	13.98	5.93	9.79 – 18.17
F_{AY} (N)	36.97	51.25	0.73 – 73.21
F_{AZ} (N)	350.39	169.46	230.57 - 470.22
M_{AX} (N*m)	22.27	10.89	14.56 – 29.97
M_{AY} (N*m)	18.91	3.34	16.54 - 21.27
M_{AZ} (N*m)	7.64	9.74	0.75 - 14.53
F_{TX} (N)	489.81	376.81	223.37 - 756.25
F_{TY} (N)	194.47	139.91	95.54 - 293.40
F_{TZ} (N)	4452.34	1834.10	3155.44 - 5749.24
F_{HX} (N)	485.62	396.57	205.20 - 766.04
F_{HY} (N)	0.00	0.00	0
F_{HZ} (N)	4284.54	1749.81	3047.24 - 5521.84
R_{OX} (N)	13.98	5.93	9.79 - 18.17
R_{OY} (N)	36.97	51.25	0.73 - 73.21
R_{OZ} (N)	259.10	126.88	169.39 - 348.82
M_{OX} (N*m)	0.00	0.00	0
M_{OY} (N*m)	18.84	3.63	16.27 - 21.41
M_{OZ} (N*m)	7.80	10.39	0.46 - 15.14

Table 1C: Measured and calculated external forces and moments in Osteoporotic femurs

Osteoporotic	Average	Standard Deviation	Range
F_{AX} (N)	3.14	4.66	0.34 - 8.52
F_{AY} (N)	112.87	21.93	91.13 - 134.99
F_{AZ} (N)	127.67	50.02	84.22 - 182.35
M_{AX} (N*m)	4.55	1.80	3.09 - 6.56
M_{AY} (N*m)	22.60	4.65	19.06 - 27.87
M_{AZ} (N*m)	5.47	4.29	2.08- 10.29
F_{TX} (N)	572.08	743.82	134.55 - 1430.93
F_{TY} (N)	87.50	65.55	48.69 - 163.18
F_{TZ} (N)	2121.78	654.81	1716.17 - 2877.20
F_{HX} (N)	569.55	738.65	134.88 - 1422.41
F_{HY} (N)	0.00	0.00	0
F_{HZ} (N)	2031.32	615.66	1717.90 - 2740.64
R_{OX} (N)	3.14	4.66	0.57 - 8.52
R_{OY} (N)	112.87	21.93	91.13 - 134.99
R_{OZ} (N)	109.06	46.06	69.14 - 159.46
M_{OX} (N*m)	0.00	0.00	0
M_{OY} (N*m)	22.50	4.77	18.77 - 27.87
M_{OZ} (N*m)	5.15	3.73	2.09 - 9.30

Table 1D: Measured and calculated external forces in Reinforced Osteoporotic femurs

Reinforced	Average	Standard Deviation	Range
F_{AX} (N)	13.52	4.41	8.95 - 17.76
F_{AY} (N)	132.72	99.93	45.96 - 241.98
F_{AZ} (N)	394.45	223.59	6.35 - 193.58
M_{AX} (N*m)	20.66	2.86	17.36 - 22.49
M_{AY} (N*m)	34.57	14.92	20.04 - 49.85
M_{AZ} (N*m)	19.31	11.87	5.61 - 26.46
F_{TX} (N)	942.94	573.60	422.80 - 1558.12
F_{TY} (N)	64.16	39.04	19.19 - 89.38
F_{TZ} (N)	5642.27	2640.93	3881.62 - 8678.88
F_{HX} (N)	950.48	583.70	413.84 - 1571.96
F_{HY} (N)	0.00	0.00	0
F_{HZ} (N)	5254.06	2352.64	3702.26 - 7960.98
R_{OX} (N)	13.52	4.41	8.95 - 17.76
R_{OY} (N)	132.72	99.93	45.96 - 241.91
R_{OZ} (N)	391.33	254.79	186.47 - 676.63
M_{OX} (N*m)	0.00	0.00	0
M_{OY} (N*m)	34.81	15.02	19.91 - 49.94
M_{OZ} (N*m)	20.05	13.02	5.03 - 28.03

DISCUSSION

The reinforcement with PMMA increased the strength of osteoporotic femurs to values comparable to the strength of normal femurs. This proved that surgical ways to reinforce osteoporotic femurs and restore strength are attainable. In all cases, it was found that the reinforced osteoporotic femurs could withstand much larger forces and moments during testing than the osteopenic or osteoporotic femurs.

The analysis of femoral forces just before fracture allowed us to find all the reaction forces and moments. Using measurements from the three load cells, we solved simplified systems of equilibrium equations. Because for some of the unknowns there was more than one equation available we solved those equations in all possible ways including using least squares. This was done in order to combine the measurements and the equations in a manner that gave us confidence in the design of our test system and the reliability of the measured data.

ACKNOWLEDGMENTS

We would like to thank the Musculoskeletal Transplant Foundation (Edison, NJ, USA) for providing the cadaveric femurs.

REFERENCES

[1] S. Cummings, M. Nevitt, and W. Browner, Risk factors for hip fractures in white women, Study of Osteoporotic Fractures Research Group, N. Engl. J. Med., Vol 332(12), 1995, p 767- 773

[2] S. Cummings, D. Bates, and D. Black, Clinical use of bone densitometry: scientific review, JAMA. Vol 288(15), 2002, p 1889-1897

[3] O. Johnell, J. Kanis, A. Oden, et. al., Predictive value of BMD for hip and other fractures. J. Bone Miner. Res., Vol 20(7), 2005, p 1185-1194

[4] T. Youm, K. Koval, and J. Zuckerman, The economic impact of geriatric hip fractures, Am. J. Orthop., Vol 28(7), 1999, p 423-428, Review

[5] A. Tosteson, S. Gabriel, M. Grove, et. al., Impact of hip and vertebral fractures on qualityadjusted life years, Osteoporos. Int., Vol 12(12), 2001, p 1042-1049

[6] S. Gabriel, A. Tosteson, C. Leibson, et. al., Direct medical costs attributable to osteoporotic fractures, Osteoporos. Int., Vol 13(4), 2002, p 323-330

[7] L. Melton, 3rd, Epidemiology of hip fractures: implications of the exponential increase with age, Bone, Vol 18(3 Suppl), 1996, p 121S-125S, Review

[8] S. Cummings, D. Black, D. Thompson, et. al., Effect of alendronate on risk of fracture in women with low bone density but without vertebral fractures: results from the Fracture Intervention Trial, JAMA, Vol 280(24), 1998, p 2077-2082

[9] D. Black, S. Cummings, D. Karpf, et. al., Randomized trial of effect of alendronate on risk of fracture in women with existing vertebral fractures, Fracture Intervention Trial Research Group, Lancet, Vol 348(9041), 1996, p 1535-1541

[10] H. Pols, D. Felsenberg, D. Hanley, et. al., Multinational, placebo-controlled, randomized trial of the effects of alendronate on bone density and fracture risk in postmenopausal women with low bone mass: results of the FOSIT study, Fosamax International Trial Study Group, Osteoporos. Int., Vol 9(5), p 461-468

[11] M. McClung, P. Geusens, P. Miller, et. al., Effect of risedronate on the risk of hip fracture in elderly women, Hip Intervention Program Study Group, N. Engl. J. Med., Vol 344(5), 2001, p 333-340

[12] A. Hodsman, L. Fraher, and P. Watson, A randomized controlled trial to compare the efficacy of cyclical parathyroid hormone versus cyclical parathyroid hormone and sequential calcitonin to improve bone mass in postmenopausal women with osteoporosis, J. Clin. Endocrinol. Metab., Vol 82(2), 1997, p 620-628

[13] R. Neer, C. Arnaud, J. Zanchetta, et. al., Effect of parathyroid hormone (1-34) on fractures and bone mineral density in postmenopausal women with osetoporosis. N. Engl. J. Med., Vol 344(19), 2001, p 1434-1441

[14] L. Melton, 3rd, Epidemiology of hip fractures: implications of the exponential increase with age, Bone, Vol 18(3 Suppl), 1996, p 121S-125S, Review

[15] S. Cummings, S. Rubin, D. Black, The future of hip fractures in the United States, Numbers, costs, and potential effects of postmenopausal estrogen, Clin. Orth. and Rel. Res., Vol 252, 1990, p 163–166

[16] J. Kanis, Assessment of fracture risk and its application to screening for postmenopausal osteoporosis: synopsis of a WHO report, WHO Study Group, Osteoporos Int., Vol 4(6), 1994, p 368-381

[17] A. Courtney, E. Wachtel, E. Myers, and W. Hayes, Effects of loading rate on strength of the proximal femur. Calcif. Tissue Int., Vol 55(1), 1994, p 53-58

DEVELOPMENT, SYNTHESIS AND CHARACTERIZATION OF POROUS BIOMATERIAL SCAFFOLDS FOR TISSUE ENGINEERING

Kajal K. Mallick
School of Engineering
University of Warwick
Coventry CV4 7AL
United Kingdom

ABSTRACT

This paper describes recent progress towards the development of freeze casting techniques using variation of camphene (CFC) and water and glycerol (WGFC) to fabricate highly porous network structures of Bioglass, hydroxyapatite (HAP), tricalcium phosphates (TCP), as well as their composites for applications in bone tissue engineering. Solid loading of 10 to 60% of ball milled ceramic slurries were freeze cast followed by sublimation at -70 to 60°C. The green bodies equilibrated at room temperature were then sintered in air to a maximum temperature of 1100°C for 4h, which produced excellent defect free and three-dimensionally interconnected structures with open micro and macropores. The nature of the pore channels varied from dendritic to lamellar geometry with a maximum size of ~ 100μm and the uniform strut thickness varying from 1-8μm. The porosity of the bioscaffolds increased monotonically from 10 to 70% with the decrease in loading volume. Microstructural analysis of porous structures was used to extract information on geometry, porosity and pore size distribution. DTA-TGA, SEM, XRD and density were used to characterize precursor powders, slurry and sintered products. The study proved particularly useful to assess the suitability of the 3-D structures produced by these methods for tissue engineering applications.

INTRODUCTION

The advent of tissue engineering has inspired many researchers to develop various synthesis methods to fabricate structures that are suitable for their use in bone tissue engineering[1]. This approach of bone augmentation avoids the problems of adverse immune response, donor site morbidity and pathogen disease transmission commonly associated with autograft/allograft and xenograft. For this purpose, a 3-D skeletal structure is considered ideal for aiding neovascularisation, osteoblasts proliferation and differentiation necessary for bone tissue generation[2-4]. Such a porous bioscaffold, using hydroxyapatite and various derivatives of calcium phospahates and bioglass[5,6], has the advantages of engineering and optimization of material composition and, in particular, porosity necessary to avoid osteolysis. Although the porosity ranges (macroporous and microporous) cited in literature vary widely from 50-1000μm[7-10] pores > 50 μm have been used successfully in bone graft substitute materials (BGS)[11-15]. Conventional ceramic fabrication methods[16-18] in producing porous structures are relatively expensive and are known to contribute to problems of binder removal and poor mechanical properties. Recently however, alternative methods of freeze casting have received considerable attention in producing interconnected structures and in-vivo studies have shown that microporosity can enhance bone growth[19,20]. The motivation for the present work is therefore to explore and develop several novel freeze casting methods that are inexpensive and can be adapted also for high solid loading. Biomaterial selection including biocomposite compositions has been based on the assumption that the degree of bioresorption can be tailored to rate of cell growth especially where rapid healing is desirable for specific clinical applications. The primary objective is to investigate relative merits of two such techniques developed in the present work namely, camphene freeze casting (CFC) and water and glycerol freeze casting (WGFC) in terms of their ability to produce networked 3-D porous structures using Bioglass, HAP, TCP and their composites. A comparative analysis of the nature and influence of the binding phase, pore structure, porosity and solid loading for the fabricated interconnected bioscaffold structures is also presented.

EXPERIMENTAL

Materials and Slurry Preparation

Commercial purity (99.9%) HAP, $Ca_{10}(PO_4)_6(OH)_2$, TCP, $Ca_3(PO_4)_2$, powders (Merck, Darmstadt, Germany) with a median particle size (d_{50}) of 3μm and a specific surface area of 72 m^2/g and the melt-derived Bioglass 45S5 powder with a particle size of <2μm were used. Camphene, $C_{10}H_6$ (Sigma-Aldrich, UK), carboxylic acid (alkali-free and molecular mass of 320 g/mol Dolapix CE64, Zschimmer and Schwarz, Germany) and glycerol (Sigma-Aldrich, UK) were used as freezing vehicle, dispersant and cryoprotectant, respectively.

Ceramic slurries were obtained by mixing precursor mixtures with a Dolapix concentration of 6 wt% followed by ball milling for 48 hours in a sealed PTFE jar. The slurry was then cast in a pre-cooled (at 0°C) brass split mould, evacuated for 20 minutes to remove any trapped air bubbles and stored in a vacuum desiccator at 0°C for 24 h followed by further processing according to the particular freeze casting method employed.

Scaffold Fabrication by Freeze Casting

A simplified flow chart of the freeze casting is shown in Figure 1. Two freeze casting methods, namely, camphene freeze casting (CFC) and glycerol and water freeze casting (GWFC) were used to fabricate scaffolds of HAP, Bioglass, and the composites of Bioglass and HAP and HAP and TCP. After careful removal from the mould followed by relevant freeze drying processing the green body was fired up to a maximum temperature of 1100°C.

Camphene Freeze Casting (CFC)

A batch weight of 40g slurry consisted solid loadings of 20, 30, 40, 50, 60 wt% containing the melt derived Bioglass, a mixture of 70:30 ratio of Bioglass-HAP, 60:40 ratio of HAP-TCP, and camphene was added as the remainder. Ball milling at a constant temperature of 60°C was carried out to maintain homogeneity and reduce the risk of the camphene solidifying. The demoulded porous green body was freeze dried between -20 to -70°C, sintered at 5-10°C/min to a maximum temperature of 1100°C (±2°C) dwelling for 2-4h and was finally cooled at the same rate to room temperature.

Water and Glycerol Freeze Casting (WGFC)

30g batch weight of a fixed 40 wt% solid HAP loading was used to produce ball milled slurry. The Glycerol content was 0, 10, 20, 30 and 40 wt% with water as the remainder. The mould, frozen by liquid nitrogen for 30 minutes, was placed in a thermally insulated polystyrene container thus forming the initial green body after casting. The porous castings in the mould still housed in the polystyrene box were freeze dried for sublimation and subsequently sintered at 1100°C for 4h at a ramp rate of 10°C/min

Thermal, Microstructural, Phase and Pore Analysis

Isothermal behavior of as-prepared slurries was determined by simultaneous differential thermal analysis and thermogravimetry (DTA-TGA) (STA1500 TA Instruments, West Sussex, UK). 20 mg of slurry was heated to 1000°C with a ramp rate of 10°C/min in flowing air.

The crystalline phase compositions of porous scaffolds sintered at various temperatures was determined using powder X-ray diffraction (XRD) in the region of $2\theta = 10\text{-}80°$ with a step size of $0.02°$ and step duration of 0.5s on a Ni-filtered Philips diffractometer (Model PW1710) using CuK$_\alpha$

radiation (λ = 0.15406 nm). Using an automated powder diffraction software package the evolved phases were matched to both standard ICDD and calculated ICSD diffraction files. A Philips Cambridge Stereoscan and JEOL Model 840 were used to characterize pore morphology as well as to observe related microstructural features of the porous scaffold structures.

Open porosity and bulk density of the green body and the as-fired scaffolds were determined using the well known Archimedes method, pyconometry (Model AccuPyc II 1340, Micromeretics, UK), and mercury porosimetry (Autopore IV 9500, Micromeretics, UK). Depending on the density of biomaterial used the porosity p of the scaffold was calculated by

$$p = 1 - \frac{\rho_{Scaffold}}{\rho_{Solid}}$$

where ρ_{solid} = 3156 kgm^{-3} for HAP; 2700 kgm^{-3} for Bioglass; 3120 kgm^{-3} for β-TCP

RESULTS AND DISCUSSION

Thermal Behavior

DTA-TGA thermograms indicated camphene melting at 65°C and the slurry admixed with DMC and Dolapix is completely burnt off at around 300°C. An endotherm at 590°C corresponded to softening point/glass transition (T_g) and an exotherm at 730°C related to the formation of crystal phase (T_c), respectively via slow conversion from glass to glass-ceramic. Melting (T_m) was observed at around 1050°C. These events are well known and have been reported in the literature[21-23].

Phase Evolution and Crystallization

XRD traces of the Bioglass, HAP, TCP, Bioglass-HAP and HAP-TCP composite powders heated at various temperatures showed that both as fabricated HAP and TCP powders are phase pure and match the standard JCPDS reference patterns (72-1243/9-432 and 9-169, respectively). The analysis of the ceramic HAP-TCP composite indicated the relative intensity ratios of the constituent phases, even taking the mass absorption coefficient in consideration, approximated the actual stoichiometric mixture of the composites.

Crystallization behavior of the as-melted Bioglass sample showed a broad peak characteristic of the absence of short range atomic order observed in amorphous materials. At higher sintering temperatures of 850-100°C, thermally induced surface crystallization to a glass-ceramic occurred as evidenced by corresponding increase in peak intensity. The crystalline phase, identified as $Na_2Ca_2Si_3O_9$ and reported in previous studies[24-28], matches the standard JCPDS reference pattern (22-1455). The effect of a crystallized phase on bioactivity of a glass-ceramic depends on several key factors such as degree of crystallization and glass composition. Formation and kinetics of hydroxyl calcium apatite (HCA) serves also as an indicator of bioactivity. There are conflicting reports surrounding the effect of crystallization of Bioglass 45S5 on bioactivity from the bioglass transforming to an inert material[29] to having little or no adverse effect on the bioactivity except where surface reaction kinetics slow down only when degree of crystallization exceeded 60%[26,30]. It is also shown that whilst the presence of $Na_2Ca_2Si_3O_9$ phase depressed the kinetics the formation of HCA layer was not suppressed[24,25,30-32]. In the present work, as the crystallization is limited to surface with the remainder being mainly amorphous glassy phase the level of bioactivity is expected to be still adequate for resorption into bone in-growth.

Bioglass-HAP composite powders showed the structure to be primarily a mixture of crystalline HAP and an amorphous phase, which on sintering at higher temperatures transformed into glass-ceramic composites. The presence of HAP in the precursor mixture was found to suppress the extent of glassy phase transforming into the glass-ceramic.

Camphene Freeze Casting (CFC)

Pore architecture

Microstructural development of porosity and pore geometry in the bioscaffolds fabricated is shown in typical SEM micrographs in Figures 2-4 for Bioglass and Bioglass-HAP composites, respectively. In all cases, 3-D interconnected micropore constructs were achieved with dendritic pore morphology that resulted from the replication of unidirectional camphene sublimation. Regular coralline nature of pores and the struts (Figure 2a-c) is facilitated by controlled and uniform viscous flow at 850°C of the finely divided 2μm glass frit. The general porous architecture remained relatively unaltered even when the solid loading was as high as 60%. Some pores were observed near or above 100μm compared to the measured mean pore size of 65μm. No other studies exist in the literature for porous Bioglass and Bioglass composite scaffolds fabricated using camphene method although a related work[27] has been reported on replicated macroporous bioglass foams.

Effect of sintering temperature

The relationship between the sintering process and the development of the porous network in Bioglass derived scaffolds is illustrated in Figure 3 for 20wt% solid loading. When sintered at 850-950°C for only 2h, the Bioglass converted to glass-ceramic via surface crystallization of $Na_2Ca_2Si_3O_9$. It is reported that crystallization influences bioactivity of such a glass-ceramic and facilitates bone regeneration[24,32]. According to the thermal data the viscous flow, at a lower temperature of 850°C, is expected to be low and the thus the pore structure is still intact. This observed minimal degree of surface crystallization behavior, reported elsewhere[33], is confirmed by still a broad XRD trace and therefore glassy. At an intermediate temperature of 950°C, the Bioglass now becomes nominally more viscous that resulted in the pore walls becoming denser and the network interconnectivity more delineated (Figure 3a). At 1050°C, close to the onset of glass melting temperature, the porous network collapsed due to substantial viscous flow of the glass-ceramic walls. Figure 3b shows the melting of glass along with the crystallized layer in the form of depressed areas where pores existed previously. Therefore, an optimum sintering temperature of 850-950°C for fabricating such glass-ceramic scaffolds is preferable in order to generate an arrayed interconnected porous network structure. Since these bioactive glasses are known to elicit cell proliferation, such constructs when implanted should result in better bone in-growth that can be tailored to the rate of biodegradation i.e. via controlled biodegradability[33-35] of bioglass and related glass-ceramics. Various supporting reports are available in the literature[36-40]. Also, the compressive strength is estimated to be around 0.3-0.7 MPa[27,28] due to the absence of any significant defects in these samples.

Figure 4 shows typical microstructure for Bioglass-HAP composite scaffolds for 20% loading. Since the composites contain predominantly ceramic hydroxyapatite (70 wt%) compared to Bioglass (30 wt%), viscous flow at 850°C is limited by the ceramic particulates with a very high melting temperature. At 950°C the pore channels, produced originally by the directional sublimation of camphene, are now more delineated. In contrast to the case for Bioglass only, the scaffold structure at 1050°C does not collapse as HAP particles now well supporting the porous network even though melting of the Bioglass is already underway. The thickness of pore walls varied from 8-10μm producing networked pores, thought to be formed around the HAP particles via glass melting. The reproducibility of the network structure would therefore depend on two interlinked factors i.e. precursor particulate size and heat treatment temperature. For Bioglass, it was observed that the overalls porosity is usually about 20-30% lower for starting particle size > 30μm compared to a size range of 1-4μm. This is supported by previous studies on calcium phosphate glass[41] where the difference in porosity can be as high as 10% between fine (1-5μm) and large particle size (15-20μm). Clearly, for fabrication, an optimum sintering temperature of up to 1000°C can be used especially for a lower solid loading and narrow size distribution at the expense of perhaps a slight reduction in porosity.

Influence of solid loading on porosity

Porosity of the scaffolds varied linearly with the solid loading of 20-60% (Figure 5). A maximum porosity of nearly 73% for Bioglass and only 65% for composites (due to limited viscous flow) were achieved. An almost uniform strut thickness for the lowest solid loading (20%) was an average of 5μm which increased with loading level as the porosity decreased. There was an increase in number of dendritic bridges/struts between the pores whilst pores became smaller as the solid loading increased. The trend in the linear variation of the mean pore size with loading is shown in Figure 6. At 950°C, for 20% loading the mean pore sizes were 50μm and 60μm for Bioglass and its composites, respectively but resulted in a much reduced level of microporosity of around 10μm at 60% solid loading. Apart from the increased glass viscosity a difference in porosity of nearly 10% is attributed to the variation of cooling rate i.e. temperature decrease in unit time during sublimation. The overall processing protocol is empirical and therefore there may well be a temperature differential in the liquid nitrogen pre-cooled moulds between the two experimental groups. As a result, the moulds equilibrated at different rates to the final freeze drying temperature. Also, the mechanisms of vapor transport and mass transport behavior during sublimation[42] for the Bioglass and the composites must have contributed to different freeze drying rates due, be it unavoidable, to different conditions of cooling. It is important to note here that the mean pore diameter increased with the increase in sintering temperature as the glass started to flow and produce a more delineated and highly interconnected 3-D porous structure. This is true except for Bioglass scaffolds where the network collapses at 1050°C.

HAP-TCP composite microstructure at different solid loading and fabricated at 1100 °C (Figure 7) can be described typically as cellular-type and networked. There is some evidence of localized channel/lamellar structures, a feature not observed in samples of Bioglass and its composites with growing direction parallel to the solidification direction. This is explained by the vapor and mass transport mechanism during sublimation being relatively more stable thermodynamically in these ceramic-ceramic composites than for the ceramic-glass composites. These constructs also show clear evidence of microporosity on the column walls with variation in pore diameters (20-40μm) and microporous pits of ~ 1-2μm (Figure 7b). The structure at loading of as high as 60 wt%, is less porous due to densification but still partially cellular. For all of the loadings, porosity and pore size increased linearly with decreasing loading level. The trend is very similar to those observed for Bioglass and Bioglass-HAP composites.

Water and Glycerol Freeze Casting (WGFC)

Pore structure

The effect of water and glycerol acting as a solvent vehicle on the development of pore microstructures for HAP freeze casts is shown in typical SEM micrographs in Figure 8. The fabricated structure of the sintered bodies is highly porous and interconnected and is similar to the camphene technique. A noticeable difference however is the increased regularity and homogeneity of the porous structure showing dendritic or coralline-like geometry and the formation of denser pore walls. This is shown in Figure 8 for casts with and without added glycerol. The dendritic bridges are evident throughout regardless of the solvent concentration. The walls were without noticeable defects thus increasing the size of the pores and are due to water acting as a solidification modifier which helps eliminate defects associated with the particle rejection during the freeze casting process (i.e. defects in the pore walls). Also, no voids were observed although several reports suggested that the particle rejection can result in the formation of large voids[43,44]. The lower cooling rate from liquid nitrogen temperature during casting and solvent sublimation created a reproducible but controlled pore structure.

Effect of glycerol content

The relationship of both the porosity and pore size with increasing glycerol concentration is monotonic (Figure 9). A wide variation of mean pore diameters of 10-50μm with a maximum of 55% porosity was achieved exhibiting a homogenous interconnected wall thickness of around 8μm even when the solid loading was as high as 40 wt%. The smaller pore diameter is due to the effect of glycerol which is known to reduce the size of the ice crystallites[44-46]. The pore structure, as shown in Figure 8e of a fractured section for 30 wt% glycerol concentration, with an average pore diameter of 45μm showed proliferation of interconnected microporous surface pits (2-3μm) on the dendritic bridges. Such pits have been observed previously[47] for 5-10% solid loading of HAP using camphene method. However, the reported pits were not connected as is the case in the present work. It is suggested that, on sublimation, frozen microscopic ice crystals separated by glycerol molecules formed such protrusions in certain <100> and <111> preferred growth directions[48]. Formation of such micropores is desirable and is expected to accelerate cellular response[49]. The literature on the use of water and glycerol in producing porous HAP scaffold is scarce and a recent study reported[45] constructs of only 1-2μm pores with porosity of < 50%. In contrast, it should be noted here that the present study achieved the control of microporosity for very high loadings by using hydrogen bond forming cryoprotectant. The freezing behavior was similar to those employed for aqueous slurries of alumina[44,50]. Clearly, WGFC technique demonstrates that a uniform level of microporosity can be controlled by tuning the glycerol/water content in the slurry.

It should be borne in mind that the fabricated structures are bioresorbable with varying degree and as resorption progresses the porosity is also expected to increase depending on the degree and the rate of HCA formation. This in turn is related to the choice of a preferred biomaterial and the site specificity of a particular clinical application. Therefore, the pore size range should not be regarded as absolute but would depend on complex physiological factors. The observed features of architectural patterns, in the form of dendrites/lamellae originated from the orientation along the solvent growth direction and hence porosity. This will permit directional cell growth and rapid vascularisation of the implants coupled to osteoconduction and osteoinduction characteristics. SBF experiments are ongoing to assess the validity of these assertions and to assess the biological response of these resorbable materials.

CONCLUSION

Two freeze casting techniques of camphene and WGFC with various slurry concentrations have been developed and demonstrated the effectiveness in fabricating highly porous 3-D network structures of bioresorbabale Bioglass, HAP, TCP and their composites. The results were evaluated on the basis of a comparative analysis of the pore structure of the 3-D constructs. The study has demonstrated the fabrication of bioactive Bioglass based 3-D and microporous scaffold bodies. Structures with porosity ranging from 50 to 70% have been successfully fabricated using well known CFC method. Using WGFC, Defect free and interconnected porosity of a maximum 55% was achieved for HAP based composites. In both cases, it was possible to replicate coralline-like uniform dendrites with micropores of 2-3μm on the interconnected bridges. This was found to be facilitated by very low temperatures during sublimation process. It is concluded that both methods are suitable with varying degree of control to fabricate structures which may be adapted for bone tissue engineering. Although > 50μm is an optimum range for bone in-growth microporosity may enhance in-vivo bioactivity and bone in-growth. Neovascularisation is also expected to be promoted by the presence of the observed microporosity. Some of the structures produced in the present work fall within the microporous range and are just outside the minimum macroporous values.

REFERENCES

[1]C.W. Patrick, Jr, A.G. Mikos and McIntire, Prospectus of Tissue Engineering In: Frontiers in Tissue Engineering, Editors: C.W. Patrick, Jr, A.G. Mikos and L.V. McIntire, Elsevier Science, New York, USA, 3–14 (1998).

[2]R. Langer and J.P. Vacanti, Tissue engineering, *Science*, **260**, 920–926 (1993).

[3]L.L. Hench and J.M Polak, Third-Generation Biomedical Materials, *Science*, **295** [5557] 1014-17 (2002).

[4]J.R.Jones and L.L. Hench, Regeneration of Trabecular Bone using Porous Ceramics, *Current Opinion Solid State Mater. Sci.*, **7** [4-5] 301-7 (2003).

[5]L.L. Hench, Bioceramics: From Concept to Clinic, *J. Am. Ceram. Soc.*, **74** [7] 1487-510 (1991).

[6]M. Vallet-Regi, A. Ramila, S. Padilla and B. Munoz, Bioactive Glasses as Accelerators of Apatite Bioactivity, *J. Biomed.Mater. Res.*, **66A** [3] 580-85 (2003).

[7]J.R.Jones and L.L. Hench, Regeneration of Trabecular Bone using Porous Ceramics, *Current Opinion Solid State Mater. Sci.*, **7** [4-5] 301-7 (2003).

[8]R.E. Holmes, Bone Regeneration within a Coralline Hydroxyapatite Implant, *Plast. Reconstr. Surg.*, **63** 626–33 (1979).

[9]O. Gauthier, J.M. Bouler, E. Aguado, P. Pilet and G. Daculsi, Macroporous Biphasic Calcium Phosphate Ceramics: Influence of Macropore Diameter and Macroporosity Percentages on Bone Ingrowth, *Biomaterials*, **19** 133–39 (1998).

[10]J. Bobyn, R. Piliar, H. Cameron and G. Weatherly, The Optimal Pore Size for the Fixation of Porous Surfaced Metal Implants by the Ingrowth of Bone, *Clin. Orthop. Rel. Res.*, **150** 263–70 (1980).

[11]P.S. Eggli, W. Muller and R. K. Schenk, Porous Hydroxyapatite and Tricalcium Phosphate Cylinders with Two Different Pore Size Ranges Implanted in the Cancellous Bone of Rabbits. A Comparative Histomorphometric and Histologic Study of Bony Ingrowth and Implant Substitution, *Clin. Orthop. Rel. Res.*, **232** 127-38 (1988).

[12]K.A. Hing, S.M. Best, K.E. Tanner, W. Bonfield and P.A. Revell, Mediation of Bone Ingrowth in Porous Hydroxyapatite Bone Graft Substitutes, *J. Biomed. Mater. Res.*, **68A** [1] 187-200 (2004).

[13]R. E. Holmes, V. Mooney, R. Buchholz and A. Tencer, A Coralline Hydroxyapatite Bone Graft Substitute *Clin. Orthop. Rel. Res.*, **188** 252-62 (1984).

[14]J.J. Klawitter, J.G. Bagwell, A.M. Weinstein, B.W. Sauer and J.R. Pruitt, An Evaluation of Bone Growth into Porous High Density Polyethylene, *J. Biomed. Mater. Res.*, **10** 311-23 (1976).

[15]J. H. Kuèhne, R. Bartl, B. Frish, C. Hanmer, V. Jansson and M. Zimmer, Bone Formation in Coralline Hydroxyapatite. Effects of Pore Size Studied in Rabbits, *Acta Orthop. Scand.*, **65** [3] 246-52 (1994).

[16]P. Sepulveda and J.G. P. Binner, Processing of Cellular Ceramics by Foaming and in situ Polymerization of Organic Monomers, *J. Eur. Ceram. Soc.*, **19** 2059–66 (1999).

[17]C. Tuck and J. R. G. Evans, Porous Ceramics Prepared from Aqueous Foams, *J. Mater. Sci. Lett.*, **18** 1003–5 (1996).

[18]S. Dhara and P. Bhargava, A Simple Direct Casting Route to Ceramic Foams, *J. Am. Ceram. Soc.*, **86** 1645–50 (2003).

[19]K A Hing, Characterisation of Porous Hydroxyapatite, *J. Mat. Sci.:Materials in Medicne*, **10** 135-45 (1999).

[20]K A Hing, Microporosity Enhances Bioactivity of Synthetic Bone Graft Substitutes, *J. Mat. Sci.: Materials in Medicne*, **16** 467-75 (2005).

[21]H.A. Batal, M.A. Azooz, E.M.A. Khalil, A. Soltan Monem and Y.M. Hamdy, Characterization of Some Bioglass–ceramics, *Mater. Chem. Phys.*, **80 [3]** 599–609 (2003).

[22]A. El Ghannam, E. Hamazawy and A. Yehia, *J. Biomed. Mater. Res.*, Effect of Thermal Treatment on Bioactive Glass Microstructure, Corrosion Behavior,Potential, and Protein Adsorption, **55 [3]** 387–398 (2001).

[23]X. Chatzistavrou, T. Zorba, E. Kontonasaki, K. Chrissafis, P. Koidis and K.M. Paraskevopoulos, Following Bioactive Glass Behavior Beyond Melting Temperature by Thermal and Optical Methods, *Phys. Stat. Sol.(a),* **201 [5]** 944–951 (2004).

[24]D.C. Clupper, J.J. Mecholsky Jr, G.P. LaTorre, D.C. Greenspan, Sintering Temperature Effects on the In Vitro Bioactive Response of Tape Cast and Sintered Bioactive Glass–ceramic in Tris Buffer, *J. Biomed. Mater. Res.*, **57 [4]** 532–40 (2001).

[25]D.C Clupper, J.J Mecholsky Jr., G.P LaTorre, D.C Greenspan, Bioactivity of Tape Cast and Sintered Bioactive Glass–ceramic in Simulated Body Fluid, *Biomaterials*, **23 [12]** 2599–2606 (2002).

[26]O.P. Filho, G.P. LaTorre, L.L. Hench, Effect of Crystallization on Apatite-layer Formation of Bioactive Glass 45S5, *J. Biomed. Mater. Res.*, **30** 509 (1996).

[27]Q.Z. Chen, I.R. Thomson and A.R. Boccacinni, 45S5 Bioglass® – Derived Glass-ceramic Scaffold for Bone Tissue Engineering, *Biomaterials*, **27 [11]** 2414-25 (2006).

[28]I.K. Jun, Y.H. Koh and H.E. Kim Fabrication of Highly Porous Bioactive Glass-ceramic Scaffold with High Surface Area and Strength, *J. Amer. Ceram. Soc.*, **89 [1]** 391-4 (2006).

[29]P. Li, Q. Yang, F. Zhang and T. Kokubo, The Effect of Residual Glassy Phase in a Bioactive Glass-ceramic on the Formation of its Surface Apatite Layer In Vitro, *J. Mater. Sci : Materials in Medicine*, **3 [6]** 452–56 (1992).

[30]O. Peitl, E.D. Zanotto and L.L. Hench, Highly Bioactive P_2O_5–Na_2O–CaO–SiO_2 Glass Ceramics, *J. Non-Cryst. Solids*, **292 [1-3]** 115–26 (2001).

[31]D. C. Clupper and L. L. Hench, Crystallization Kinetics of Tape Cast Bioactive Glass 45S5, *J. Non-Cryst. Solid*, **318 [1-2]** 43-8 (2003).

[32]O. Peitl, G.P. LaTorre and L.L. hench, Effect of Crystallisation on Apatite-Layer Formation of Bioactive Glass 45S5, *J. Biomed. Mater. Res.*, **30 [4]** 509-14 (1996).

[33]L.L. Hench, Sol-gel materials for bioceramic applications, *Current Opinion Solid State Mater. Sci.*, **2** 604-10 (1997).

[34]A.E. Clark and L.L. Hench, Calcium Phosphate Formation on Sol-gel Derived Bioactive Glasses *J. Biomed. Res.*, **28** 693-8 (1994).

[35]L.L. Hench and J. Wilson, Surface Active Biomaterials, *Science*, **226** 630-6 (1984).

[36]L.L. Hench, R.J Splinter, W.C Allen, T.K Greenlee, *J. Biomed. Mater. Res. Symp.*, **2 [1]** 117-41 (1971).

[37]W. Cao and L.L. Hench, Bioactive materials, *Ceram. Int.*, **22** 493–507 (1996).

[38]T. Kokubo, Novel Bioactive Materials, *Anales de Química*, **93** S49-55 (1997).

[39]V. Banchet, E. Jallot, J. Michel, L. Wortham, D. Laurent-Maquin and G. Balossier, *Surf. Interface Anal.*, X-ray Microanalysis in STEM of Short-term Physicochemical Reactions at Bioactive Glass Particle/Biological Fluid Interface. Determination of O/Si Atomic Ratios, **36 [7]** 658–65 (2004).

[40]T. Kokubo, H. Kushitani and S. Sakka, *J. Biomed. Mater. Res.*, Solutions able to Reproduce in vivo Surface-structure Changes in Bioactive Glass-ceramic A-W3, **24 [6]** 721-34 (1990).

[41]C. Wang, T. Kashihiro and M. Nogami, Macroporous Calcium Phosphate Glass-ceramic Prepared by Two-step Pressing Technique and Using Sucrose as a Pore Former, *J. Mat. Sci.:Materials in Medicine*, **16 [8]** 739-44 (2005).

[42]M. Kochs, C.H. Körber, B. Nunner, I. Heschel, The Influence of the Freezing Process on Vapour Transport during Sublimation in Vacuum Freeze Drying, *Int. J. Heat Mass Transfer*, **34 [9]** 2395-408 (1991).

[43]K.A.Keler, G.M. Mehrotra and R.J. Kernas, Freeze Forming of Alumina Monoliths: pp. 557-67, in Processing and Fabrication of Advanced Materials. V. Edited by T.S. Srivatsan and J.J. Moore, Minerals, Metals and Materials Society/AIME, Warrendale, PA, (1996).

[44]S.W. Sofie and F. Dogan, Freeze Casting of Aqueous Alumina Slurries with Glycerol, *J. Amer. Ceram. Soc.*, **84 [7]** 1459-64 (2001).

[45]Q.Fu, M.N. Rahaman, F Dogan and B.S. Bal, Freeze Casting of Porous Hydroxyapatite Scaffolds: I. Processing and General Microstructure Solutions, *J. Biomed. Mater. Res. Part B: Appl. Biomater.* (2008), DOI: 10.1002/jbm.b30997

[46]Q.Fu, M.N. Rahaman, F Dogan and B.S. Bal, Freeze Casting of Hydroxyapatite Scaffolds for Bone Tissue Engineering Applications, *Biomedical Materials*, **3** (2008) IOP Publishing, UK 1-7 (2008).

[47]B-H. Yoon, Y-H. Koh, C-S. Park, and H-E. Kim, "Generation of Large Pore Channels for Bone Tissue Engineering Using Camphene-Based Freeze Casting, *J. Am. Ceram. Soc.*, **90 [6]** 1774-752 (2007).

[48]E.R. Rubenstein and M.E. Glicksman, Dendritic Growth Kinetics and Structure II. Camphene, *J. Crys. Growth*, **112** 97-110 (1991).

[49]K. Anselme, Osteoblast Adhesion in Biomaterials, *Biomaterials*, **21 [7]** 667-81 (2000).

[50]K. Lu, C.S. Kessler and Richey M. Davis, Optimization of a Nanoparticle Suspension for Freeze Casting, *J. Am. Ceram. Soc.*, **89 [8]** 2459-65 (2006).

Presented at the MS&T'08 Conference, Pittsburgh, USA, October 4-9, 2008
[*]Member, American Ceramic Society
[**]Author to whom correspondence should be addressed
    email: k.k.mallick@warwick.ac.uk

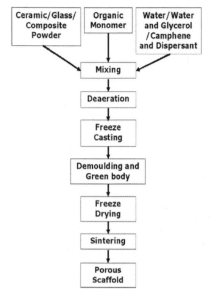

Figure 1. Flow chart of freeze casting methods for the fabrication of porous bioscaffolds

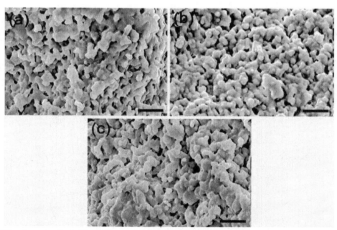

Figure 2. SEM micrographs of the CFC Bioglass scaffolds sintered at 850 °C using solid loading of (a) 20 wt%, (b) 40 wt% and (c) 60 wt% [bar = 200µm].

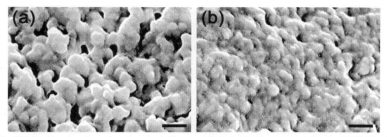

Figure 3. SEM microstructure showing the influence of sintering temperature for the CFC Bioglass scaffolds for a solid loading of 20 wt% at 950°C (b) and 1050°C (c) [bar = 100μm].

Figure 4. SEM micrograph showing the influence of sintering temperature for the CFC Bioglass-HAP composite scaffolds for a solid loading of 20 wt% at 950°C [bar = 100μm].

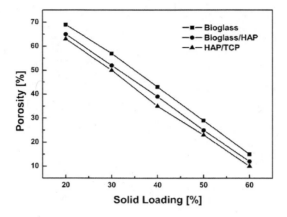

Figure 5. Effect of measured porosity with solid loading for CFC scaffolds sintered at 850°C (■ Bioglass,● Bioglass-HAP composite and ▲ HAP-TCP composite)

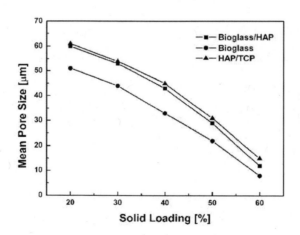

Figure 6. Plot of mean pore size of CFC scaffolds as a function of solid loading heat treated at 850°C (■ Bioglass-HAP composite,● Bioglass and ▲ HAP-TCP composite).

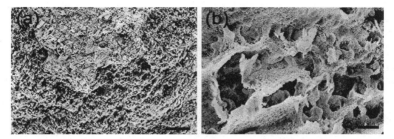

Figure 7. SEM micrographs of the porous architecture of CFC HAP-TCP composite scaffolds sintered at 1100°C for 4h at solid loading of (a) 20 wt%, and (b) 30 wt% [bar = 1mm (a); bar = 20μm (c)].

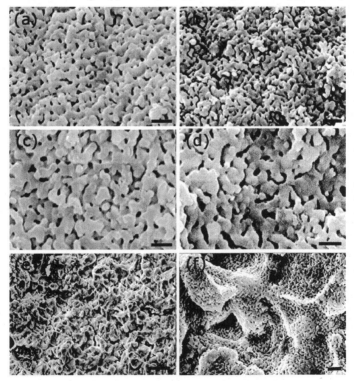

Figure 8. SEM microstructure of the WGFC HAP-TCP composite scaffolds with a solid loading of 40 wt% showing the influence of glycerol concentration in water with (a) without glycerol (b) 10 wt%, (c) 20 wt%, (d) 30 wt%, (e) atypical features at 30 wt% and (f) magnified image of (e) [bar = 200μm (a), (c), (e); bar = 100μm (b) and (d); bar = 5μm (f)].

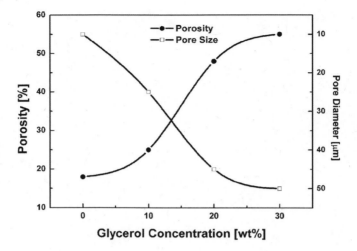

Figure 9. Influence of solvent concentration on porosity and pore size of WGFC HAP-TCP composite scaffolds at 40 wt% solid loading (● Porosity and □ Pore size).

ENGINEERED NANOFIBERS WITH STEM CELLS FOR BIOMIMETIC TISSUE ENGINEERING

Seeram Ramakrishna*, Susan Liao, Kun Ma, Casey K Chan
National University of Singapore
Singapore

ABSTRACT:
Attempts have been made to fabricate scaffolds to mimic the chemical composition and structural properties of extracellular matrix because one would think that a tissue-engineered scaffold with these characteristics will have a better chance at enhancing tissue regeneration. Nanofibers scaffold with versatile patterns have been successfully and relatively simple produced from many synthetic and natural polymers through the nanotechnology of electrospinning. Those patterns may mimic the diversity of tissue-specific orientation of fibers. Several studies have been conducted on bone marrow-mesenchymal stem cells (BM-MSCs) and bone marrow-hematopoietic stem cells (BM-HSCs) attachment on electrospun collagen nanofibers and synthetic nanofibers. After further functionalization of nanofibers by biomolecules, the nanofiber has also been developed to capture stem cells efficiently. However, the significance of certain type of nanotopography or combination of several types of nanotopography in directing the expansion and differentiation of stem cells is need to be demonstrated further in the future.

INTRODUCTION
The principal components of extracellular matrix (ECM) are structural protein: collagen, specialized proteins and proteoglycans, mainly in the form of fibers and fibrils. One specific design objective of a porous scaffold for tissue engineering is to fabricate a porous scaffold out of absorbable polymer that mimic the extracellular matrix in supporting cell proliferation and organization. In general, scaffolds used for tissue engineering should have the following characteristics: biocompatibility, biodegradability, reproducibility, high porosity with interconnected pores, and no potential of serious immunological or foreign body reactions. Owing to their functional properties and design flexibility, polymers are the primary choice of materials for making scaffolds. Polymers used for making scaffolds are classified as either naturally derived polymers or synthetic polymers. The natural polymer includes collagen, gelatin, chitosan, chitin, cellulose, and starch. The synthetic polymer includes frequently used biodegradable synthetic polymers such as poly(lactic acid) (PLA), poly(glycolic acid) (PGA), poly(lactic-co-glycolic acid) (PLGA), poly(ε-caprolactone)(PCL) and poly(lactic-co-caprolactone) (PLA-CL). These are all approved by the US Food and Drug Administration (FDA) for certain biomedical applications.

Although a number of fabrication technologies have been applied to process biodegradable and bioresorbable materials into three-dimensional (3D) polymeric scaffolds with high porosity and surface area, most of these methods focus on fabrication at the micro or macro level. Recently, nano-scaled controls of scaffold are highly popular in fabricating biomimetic scaffolds with characteristics which mimic the natural ECM. Fig.1 showed nanofibrous scaffold with different patterns fabricated by our group using nanotechnology of electrospinning. Electrospinning is an easy and versatile technique to fabricate biomimetic nanofibrous scaffolds.

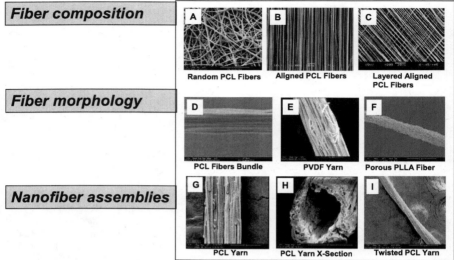

Figure 1. Electrospun nanofiber and its assemblies with various patterns.(Scales in A,C,D E are 20 μm, scale in B is 10 μm; scale in F is 50 μm, scales in G and H are 100 μm, scale in I is 200 μm)

Certain adult stem cell types are pluripotent, meaning that they can differentiate into cells derived from three germ layers. Hematopoietic stem cells (HSCs) are stem cells that give rise to all the blood cell including myeloid and lymphoid lineages. HSCs may differentiate into another three major cell types: brain cells; skeletal and cardiac muscle cells; and liver cells. The multipotent differentiation potential of human bone marrow-derived MSCs into adipocytes, osteoblasts, chondrocytes, etc has also been well characterized. However, the culture techniques of stem cells are yet to be standardized, and there are still a lot of unknown factors to be studied in the near future. The bioreactor technology has also led to the discovery of the mechanical stimulus for cell functions, which is also important for stem cell differentiation. It was found that when handling the stem cells, several growth factors or biomolecules should be supplemented in the base medium to induce differentiation into specific somatic cell types. In order to maintain the phenotype expression and differentiated functions of stem cells, the simulated natural environment of the biomimetic ECM support has to provide the appropriate signals to the attached cells. Scaffolds with nanotexture can provide physical, chemical as well as spatial cues that are essential to mimic the natural tissue growth including cell adhesion, proliferation and differentiation. In this study, we hypothesized that both topographical and biochemical cues of the substrate could promote HSCs and MSCs adhesive behaviors, which are crucial for stem cell spreading, self-renewal and lineage commitment within their microenvironment.

MATERIALS AND METHODS
1. Fabrication of nanofibers
The collagen-blended PLGA nanofiber, pure PLGA nanofiberand pure collagen nanofiber were electrospun utilizing the same methods as our study[1]. Briefly, Polymer solution

was placed in a plastic syringe and dispersed at a constant feeding rate by a syringe pump at humidity of 55 % - 60 % and temperature of 20 – 24 ˚C. High voltage of 12-15 kV DC was applied between the needle and the collector by a high-voltage power supply (Gamma High Voltage Research, Ormond Beach, FL).The electric field generated by the surface charge caused the solution drop at the tip of the needle to distort into a Taylor cone. Once the electric potential at the surface charge exceeded a critical value, the electrostatic forces would overcome the solution surface tension and a thin electrified jet of solution would erupt from the surface of the cone. The resultant nanofibers were collected on cover slips located 12 cm from the needle tip. After solvent evaporation, the nanofiber-coated cover slips were placed in a vacuum dryer overnight to remove any remaining solvent.

Scanning electron microscopy (SEM) micrographs were obtained with a JSM-5800LV scanning electron microscope (JEOL, Tokyo, Japan) to observe the morphology of nanofibers. The diameter range of the fabricated nanofibers was measured on the basis of SEM images, using image analysis software (Image J, National Institutes of Health, Bethesda, MD).

2. Modifications of nanofibers by biomolecules

Fabricated nanofiber was coated overnight at 4˚C by 10 μg / ml E-sel / IgG1 Fc chimeras in 20 mM Tris-buffered saline (PH 7.4) with 1mM $CaCl_2$. Then it was washed 3 times with Phosphate-Buffered Saline (PBS) supplemented with 0.1% BSA and 1 time with IMDM supplemented with 1% BSA.

Protein G was diluted in bicarbonate buffer (pH 9.4, Bup^{TM} Carbonate) at a concentration of 10 μg/ml and coated on the surface of blended nanofiber for 1h at room temperature. Then the nanofiber was washed by Phosphate-Buffered Saline (PBS) twice. CD29 Ab was then added onto protein G-coated NFS and incubated at 37 °C for 1h and washed by PBS twice.

3. Capture of BM-HSCs or BM-MSCs by nanofibers

0.5 ml BM-HSCs or BM-MSCs with an initial density of 20,000 cells/ml were seeded into each well of 1.88 cm^2 of 24-well plates covered with different substrates: 1.Tissue Culture Polystyrene (TCP) 2. Pure PLGA nanofiber 3. Pure Collagen nanofiber; 4. Collagen-blended PLGA nanofiber 5. Collagen-blended PLGA nanofiber coated with E-selectin or CD29 Ab respectively. The cells suspension was exposed to these substrates at room temperature for defined periods of time and washed by PBS twice to remove unattached cells. The captured cells were then observed and counted by optical microscopy and SEM micrographs. The capture percentages were calculated based on the initial seeding density.

4. Statistical analysis

Values (at least triplicate) were averaged and expressed as means ± standard deviation (SD). Each experiment was repeated three times. Statistical differences were determined by Student two-sample test. Differences were considered statistically significant at $p < 0.05$.

RESULTS AND DISCUSSIONS

On the basis of preliminary study on the stem cells capture on nanofibers, collagen nanofiber is the better choice than any other synthetic nanofiber, and is comparable to collagen coated synthetic polymer nanofiber. However, the fast degradation and weak mechanical property of collagen nanofiber prevent its applications on tissue engineering. Thus, we proposed the blended synthetic polymer with 50 % collagen to fabricate nanofibers for stem cell capture.

From Fig.2A, the blended PLGA/collagen significant enhanced the HSCs capture within 30 and 60 minutes. Capture efficiency studies showed that blended PLGA/collagen nanofibers, after being coated with E-selectin, significantly increased the HSC capture percentage from 23.40 % to 67.41 % within 30 minutes, and from 29.44 % to 70.19 % within 60 minutes of incubation at room temperature (Fig. 2B). BM-HSCs with its typical rounded morphology were captured on E-selectin coated NFS after 30 mins of incubation at room temperature (Fig. 3A).

The study with MSC showed promising results (Fig.4) too. Over 50 % percent of the MSC population adhered to the collagen nanofibers with focal adhesion (Fig.3B), while cover slip did not enhance MSCs adhesion after 30 minutes at room temperature (Fig.4A). The capture efficiency could be increased to over 70 % when the collagen nanofibers were pre-coated with stem cell specific monoclonal antibodies which recognize CD29 or CD49a antibodies. CD29 antibody linked PLGA/collagen significantly increased the MSC capture percentage from 10 to 60 minutes.

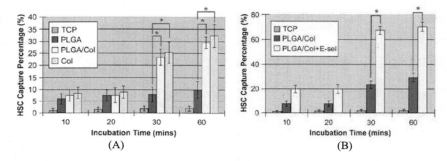

Figure 2. HSCs capture percentage on nanofibers (A) and functionalized nanofibers (B).

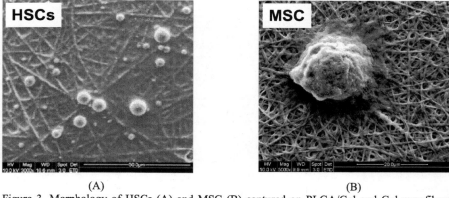

(A)   (B)

Figure 3. Morphology of HSCs (A) and MSC (B) captured on PLGA/Col and Col nanofibers respectively.

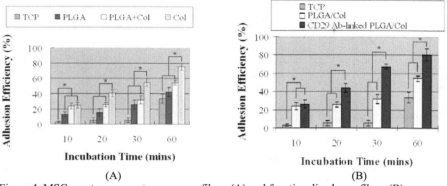

Figure 4. MSCs capture percentage on nanofibers (A) and functionalized nanofibers (B).

Polymeric nanofiber scaffolds have many desirable properties for constructing 3D niche-like matrix, such as a large surface-to-volume ratio, high porosity, ECM-like architecture, biocompatibility, availability of surface modification and flexibility of loading growth factors or drugs. In this study, we employ an easy and versatile technique of electrospinning to fabricate biomimetic nanofibers for rapid and rich capture of stem cells. The nanometer scale architecture of nanofiber scaffold resembles the natural HSCs and MSCs niche, which is composed of a 3D network of nanoscale fibrous proteins including collagen and proteoglycans.[2, 3]

CONCLUSION

Our results indicate the feasibility of promoting the adhesive characteristic of stem cells (HSCs and MSCs) by modulating the topographical and biochemical properties of the culture substrate. This study suggests the great potential for designing biomimetic artificial nanofibers for tissue engineering applications as efficient anchorage for stem cells to facilitate expansion or differentiation functions.

ACKNOWLEDGEMENT
This study was supported by MDRP and R397000036112, National University of Singapore, MOE.

REFERENCES
[1]K. Ma, C.K. Chan, S. Liao, W.Y.K. Hwang, Q. Feng, S. Ramakrishna. Engineered Electrospun Nanofiber Scaffolds for Fast and Rich Capture of Bone Marrow-Derived Hematopoietic Stem Cells. *Biomaterials* **29**:2096-2103 (2008).
[2]K.N. Chua, C. Chai, P.C. Lee, Y.N. Tang, S. Ramakrishna, K.W. Leong, H.Q. Mao. Surface-aminated electrospun nanofibers enhance adhesion and expansion of human umbilical cord blood hematopoietic stem/progenitor cells. *Biomaterials* **27**:6043-6051 (2006).
[3]C.P. Barnes , S.A. Sell, E.D. Boland, D.G. Simpson, G.L. Bowlin. Nanofiber technology: Designing the next generation of tissue engineering scaffolds. *Adv Drug Deliv Rev* **59**:1413-1433 (2007).

PRECLINIC TEST OF COLLAGEN MEMBRANES

B. León Mancilla[1], C. Piña Barba[2]

[1] Universidad Nacional Autonoma de Mexico, Facultad de Medicina, Depto.de Cirugia. Ciudad Universitaria, Circuito Interior s/n. C.P. 04510. México D.F. Mexico.
[2] Universidad Nacional Autónoma de México, Instituto de Investigaciones en Materiales. Ciudad Universitaria, Circuito Exterior s/n. C.P. 04510. México D.F. Mexico

ABSTRACT

Biocompatibility tests of materials proposed for medical applications are essential for determining its ability to be accepted in the organism which will be implanted. At the present time, the studies needed between others are in vitro (cell) and in vivo: preclinical (animal) and clinical (human) tests.

In this work were used porous membranes of collagen (type I) from inorganic bovine bone Nukbone®, to evaluate their biocompatibility in preclinical phase, using as animal model: rabbits.

The collagen porous membranes were tested in the back of rabbits, to which were withdrew a circle of skin where were implanted the membranes. It was found that the repair tissue process was faster and better than in other case, furthermore the new skin showed a better quality. Collagen porous membranes from Nukbone® were biocompatible.

INTRODUCTION

Collagen is a multifunctional family of proteins of unique structural characteristics and probably the most abundant animal protein in nature. It is estimated that collagen accounts for about 30 % of the total human body protein[1]. Collagen contributes to mechanical properties and tissue integrity[2]. Collagen is located in the extracellular matrix of connective tissues.

Currently known 25 different types of collagen, which differ only by their number and sequence of amino acids, collagen type I being the most abundant, as is found in bone, skin, tendon, among others and is the same for all mammals . This feature is one that interests us take to find a biocompatible material, readily available, high purity, good mechanical properties, adaptable to situations surgical and medical required to solve problems such as loss of skin burns or trauma, used as mesh in case of hernias, like separation between organs and tissues, increased volume in connective tissue, a guide for regeneration of cartilage, or like haemostatic[3-6] sponges among others.

The collagen may come from different sources and from different organs and tissues (bone, tendon, skin, etc.) from fish (i.e. salmon, shark) and from mammals, in particular from domestic mammals like bovine and pigs[7-9].

This paper aims to assess the biofuncionality of the collagen porous membranes obtained from the bone matrix of bovine bone named Nukbone®[10], in rabbits with skin lesions at different times of evolution.

The collagen membrane was used as tissue support to reduce the time repair of damaged and avoid dehydration and infection of the tissue, which was observed at different times of evolution in experimental animal model (preclinical study phase).

Up to the present there are different presentations of collagen that can be purchased commercially like: Fibrogen, Bioderm, Tegaderm, etc. [11-13].

MATERIAL AND METHODS.

The obtaining of collagen sponges were made from porous bone of mammal to which it withdrew all organic matter that surrounds the matrix of hydroxyapatite by physical chemical methods, using heat, boiling with proteases and ionic detergents, rinsing with deionized water and using centrifugation and finally ultrasound with ethyl alcohol to 100%. The bone matrix thus obtained preserves completely the bone structure, which consists of collagen covered with hydroxyapatite.

This preferably bone is immersed in an acid solution between 0.01 M and 3.0 M. The time that the bone should happen in this solution depends on its size and shape, the demineralization ends when pressing the material and then releases it this presents an elastic deformation similar to a sponge. Once the bone is dematerialized is removed from the acid solution and is washed with a phosphate buffer solution within a pH between 5 and 8. The collagen type I sponges are kept in refrigeration.

The samples were characterized by SEM and Western blot. To observe the microstructure of the samples, these were covered with gold and was used a scanning electron microscope Leica Cambridge Stereoscan 440 using a 15 kV voltage[14].

Western blot was used to confirm the type of collagen present in the samples. The proteins separated by electrophoresis were then transferred to a nitrocellulose membrane which was incubated in 1:1000 diluted in primary commercial anti-collagen I and III goat antibody (Santa Cruz,CA). After washing the membranes were incubated with second antibody for 1h[15]. Bands were visualized with diaminobenzidine. Commercial type I collagen from calf skin (Sigma- Aldrich) was used like reference.

Animal model

Were used 20 healthy male rabbits, bred of New Zealand with a weight of 2.500 to 3.500Kg. The animals remained housed in individual cages with water and food ad libitum, with controlled temperature of 21°C, relative humidity of 50-70%, and periods of 12:12 hours light / darkness. The procedures and handling of animals were carried out under Norma Oficial Mexicana ZOO-062-1999 (NOM)[16].

Surgical procedure

It was performed a trichotomy of the dorsal region of the animal and were made antiseptic procedure in the area, covering it with sterile fields. It was made a circular incision of 20 mm of diameter in the animal's spine covering the three layers of the skin 2-3 mm of depth, see Figure 1.

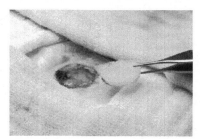

Fig. 1 Incision of the sink and collocation of membrane.

The protocol was used anesthetic ketamine 25 mg/Kg IM, xylazine 5 mg/Kg IM Pentobarbital sodium and 20 mg/Kg IV[17]. After the surgical procedure was used analgesic Dipyrone Vet 500 mg/kg every 24 h/2 days IM (trade name and laboratory in all cases).

Before his placement, the collagen membrane was hydrated with saline solution to 0.9% in order to obtain greater flexibility, see Figure 2 and then was placed on the injury, because their properties haemostatic, it was not necessary to remove or clean hematopoietic tissue from the wound, see Figure 3. To contain the membrane in its place, was necessary sutured it in four sites using Polypropylene (Prolene, Ethicon) 5/0 as shown in Figure 4.

The times of assessment for the membrane were 7, 14, 30 and 45 days after surgery. For the seizure of regenerated tissue, the animals were sacrificed with an overdose of pentobarbital (60 mg kg).

Figure 2. The membrane has the ability to adapt to the wound.

Figure 3. It is possible observe the haemo static ability of the collagen membrane.

Figure 4. Sutured membrane can be seen over the wound.

RESULTS

The animal model used for evaluating the biofuncionality of collagen membrane was acceptable, the implantation did not move at all.

The treatment, the surgical procedure and anesthetic used, were easily applicable in this specie. There was no type of infection.

It was not necessary use dressings or meshes to cover the membrane, because the biomaterial was designed according to the skin lesion leaving the membrane attached perfectly with the skin of the animal, not allowing the subcutaneous tissue get in contact with the outside world, that could leave it expose to an invasion of microorganisms and to a possible infection.

One advantage of using the dorsal skin of the animals to place the membranes is facilitating the clinical observations without any treatment to the animal that could move the membrane.

Before surgery, in all cases, the membrane was hydrated with sterilized saline solution in order to keep manageable, because when the membrane is not wet, it is very rigid. Once hydrated the collagen membrane, this kept their mechanical properties (resistance), there was not ruptured of the membrane at the time of placing the suture between the membrane and the skin of the animal. This assured the permanence of the membrane on the site of implantation.

There were not infection, nor chronic inflammation, nor detachment or rupture of the sutured membranes in any case, during the evaluation time.

The membrane permeability was observed until 72 hours after implantation, (Fig.5) during this time the membrane remained impermeable due to growth of epithelial tissue of the injury, which was promoted by the use of membrane that functioned as scaffolding.

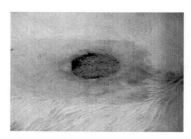

Figure 5. It can see the membrane implanted after 72 h after surgery.

There was not a contraction of the wound in any case, which would be a normal phenomenon especially for circular injuries that takes 30% more time to heal than any other geometric shape of injury.

During times of evaluation of the membranes, these remained in their place; these biomaterials retained their architecture but gradually became dehydrated.

CONCLUSIONS

The type I collagen is a protein common in all mammals, their compatibility has been proven in many published works.

This study showed that the collagen sponge obtained from the bone matrix from bovine bone presents very good functionality to establish as a substitute for skin, because the animal remained healthy and assisted in the regeneration of the skin over it. The animal model proposed for evaluating the collagen membrane was appropriate and easy to handle.

The biomaterial proposed in this work has the potential to be used in different medical procedures requiring the use of scaffolding or support cell (guided tissue regeneration) as well as transient skin covered in burns, skin ulcers or in specialties such as dentistry, plastic and reconstructive surgery, orthopedics, urology, vascular surgery, tissue engineering, etc.

ACKNOWLEDGEMENTS

To DGAPA-UNAM for financial support trough project IN104008. To Cathedra Dr. Aniceto Orantes Suárez for financial support. To Biocriss S.A. for provide all the Nukbone® necessary for this work.

To Dr. Fernando Villegas, M.en C. Ariana Labastida, Biol. Karla Dávalos, MVZ Andrés Montiel Rodriguez, Enf. Carolina Baños Galeana, Q.Carmen Peña Jiménez and Ing. Jorge García Loya by their support.

REFERENCES

[1] Ringe J, Kaps C, Burmester GR, Sittinger M. Stem cells for regenerative medicine: Advances in the engineering of tissues and organs. Naturwissenschaften (2002); 89: 338-351.

[2] Shu-Tung, L. In Biomaterials principles and applications pp. 117-139(2003) CRC Press Inc., Boca Raton, Florida.

[3] Badylak SF. Xenogeneic extracellular matrix as a scaffold for tissue reconstruction. Transplant Imnunology (2004); 12: 367-377.

[4] Trasciattia S., Podestàb A, Bonarettib S., Mazzoncinia V. and Rosinia. V.S. In vitro effects of different formulations of bovine collagen on cultured human skin. Biomate-rials (1998); **19**: 897-903.

[5] Burton JL, Etherington DJ, Peachey RD. Collagen sponge for leg ulcers. J Dermatol. (1978); **99**:681-685.

[6] Wu Z, Sheng Z, Sun T, Geng M, Li J, Yao Y, Huang Z. Preparation of collagen-based materials for wound dressing. Chin Med J (Engl). (2003); **116**:419-23.

[7] Fernandes RM, Couto Neto RG, Paschoal CW, Rohling JH, Bezerra CW. Collagen films from swim bladders: preparation method and properties. Colloids Surf B Biointerfaces. (2008); **62**:17-21.

[8] K. Fujii, T. Yamagishi, T. Nagafuchi, M. Tsuji and Y. Kuboki. Biochemical properties of collagen from ligaments and periarticular tendons of the human knee. Knee Surgery, Sports Traumatology, Arthroscopy (1994); **2**: 229 – 233.

[9] Pek YS, Gao S, Arshad MS, Leck KJ, Ying JY. Porous collagen-apatite nanocomposite foams as bone regeneration scaffolds.Biomaterials (2008)in press.

[10] M.C. Piña B., N. Munguía A., R Palma C., E.Lima.Caracterización de hueso de bovino anorgánico: Nukbone. Acta Ortopédica Mexicana (2006); **20**:150-155.

[11] http://www.fibrogen.com/collagen

[12] http://www.collagen.com, www.collagen-membrane.com

[13] http://www.collagenmatrix.com/tech-matrixeng

[14] Ariana Labastida Polito. Obtención de Membranas de Colágena a partir de hueso de bovino. Tesis de Maestría. Universidad Nacional Autónoma de México. (2006).

[15]Protein Electrophoresis. Technical Manual. Amersham Biosciences. Buckingham-shire, England. www.amershamebiosciences.com

[16] Norma Oficial Mexicana ZOO-062-1999. Especificaciones técnicas para el uso, cuidado y reproducción de los animales de laboratorio. México D.F.

[17] León MB, Villegas AF. Manual de Manejo y Anestesia en el conejo como modelo quirúrgico en docencia. (2006) 2ª. Edición. Facultad de Medicina UNAM. México ISBN ISBN: 970-32-402-3

Surface Modification
of Biomaterials

POLYSILOXANE COATINGS CONTAINING TETHERED ANTIMICROBIAL MOIETIES

P. Majumdar, S. J. Stafslien, J. Daniels, E. Lee, N. Patel, N. Gubbins, C. J. Thorson, and B. J. Chisholm
Center for Nanoscale Science and Engineering, North Dakota State University, Fargo, ND USA

ABSTRACT

An array of moisture-curable polysiloxane coatings containing chemically bound or "tethered" quaternary ammonium salt moieties derived from blends of tetradecyldimethyl(3-trimethoxysilylpropyl) ammoniumchloride (C14-QAS) and octadecyldimethyl(3-trimethoxysilylpropyl) ammoniumchloride (C18-QAS) were produced and their antimicrobial properties toward *Pseudomonas aeruginosa*, *Escherichia coli*, and *Staphylococcus epidermidis* determined. The results of study showed that a coating composition derived from a 1/3 molar ratio of C14-QAS to C18-QAS and the highest molecular weight silanol-terminated polydimethylsiloxane provided broad spectrum antimicrobial activity. This coating was used to coat urinary catheters using a dip coating process and the antimicrobial properties of the coated catheters compared to that of an uncoated catheter. The results of the evaluation showed that coating of the catheters with the QAS-functional polysiloxane provided a 54%, 94%, 80%, and 89% reduction in biofilm retention toward *E. coli*, *S. epidermidis*, *P. aeruginosa*, and *Candida albicans*, respectively.

INTRODUCTION

Quaternary ammonium salts (QASs) have been known and widely used for more than half a century as contact disinfectants. Surfaces coated with QAS-containing polymers have been shown to be very effective at killing a wide range of microorganisms such as Gram-positive bacteria, Gram-negative bacteria, yeasts, and moulds.[1,2] Alkyl chain length of the QAS plays an important role in antimicrobial activity.[3,4] Structure, density, distribution of QASs, and choice of polymer matrix can affect the overall biocidal activity. Hence, a combinatorial approach using automated high-throughput experimentation is a very useful tool to accelerate the screening and down-selection of material candidates for further evaluation.[5-7] The authors have been investigating the concept of tethered QASs to polysiloxane matrices to combat marine biofouling using a high-throughput combinatorial approach. The results showed that coatings derived from C18-QAS were effective against the marine bacterium, *Cellulophaga lytica,* while coatings derived from C14-QAS were effective against the marine diatom, *Navicula incerta*.[8] For the study described in this document, a library of twelve coatings were prepared based on three different silanol terminated polydimethylsiloxanes, two QAS concentrations, and three combinations of C14-QAS and C18-QAS. The antimicrobial activity of the coatings was evaluated toward biomedically-relevant microorganisms.

EXPERIMENTAL

Materials

2,000 g./mole silanol-terminated polydimethylsiloxane (2K-PDMS), 18,000 g./mole silanol-terminated polydimethylsiloxane (18K-PDMS), 49,000 g./mole silanol-terminated polydimethylsiloxane (49K-PDMS), tetradecyldimethyl(3-trimethoxysilylpropyl) ammoniumchloride (C14-QAS), octadecyldimethyl(3-trimethoxysilylpropyl) ammoniumchloride (C18-QAS), and methyltriacetoxysilane (MeTAS) were purchased from Gelest. 1.0 M tetrabutylammoniumfluoride (TBAF) in tetrahydrofuran was obtained from Aldrich. Toluene was obtained from VWR. 4-Methyl-2-pentanone was purchased from Alfa Aesar. A silicone reference coating, DC 3140, was obtained from Dow-Corning. Stock solutions of 80 wt% PDMS 49K in toluene and 50 mmolar TBAF in 4-methyl-2-pentanone (Cat sol) were used for coating formulations while all other reagents were used as received.

Coating Preparation

An automated coating formulation system manufactured by Symyx Discovery Tools, Inc. was used to prepare the coating solutions. Materials were dispensed into 8 ml vials using a robotic pipette having interchangeable tips and mixed with a magnetic stir bar in each vial. 0.25 mL of each coating formulations was deposited into one column of a 24 well polystyrene plate modified with epoxy primed aluminum discs. Each coating plate also contained a silicone reference coating (35% by weight suspension of DC 3140 in methylisobutylketone) which was used to compare coating performance among the plates. Coatings were cured for 24 hours at room temperature, followed by an additional 24 hour treatment at 50°C. Table 1 provides the composition of each coating solution prepared.

Catheters were dip-coated with coating solutions. Coated catheters were cured by allowing them to stand for 24 hours at room temperature followed by heating for 24 hours at 50° C.

Table 1 The compositions of the coating solutions prepared. *80 weight percent solution in toluene. All values are in grams.

Coating ID	2K-PDMS	18K-PDMS	49K-PDMS*	C14-QAS	C18-QAS	MeTAS	Cat sol
1/1C14/C18-L-2KPDMS	3.50	----	----	0.08	0.09	0.53	0.53
1/1C14/C18-L-18KPDMS	----	3.5	----	0.08	0.09	0.53	0.53
1/1C14/C18-L-49KPDMS	----	----	4.38	0.08	0.09	0.53	0.53
1/3C14/C18-H-2KPDMS	3.5	----	----	0.08	0.26	0.53	0.53
1/3C14/C18-H-18KPDMS	----	3.5	----	0.08	0.26	0.53	0.53
1/3C14/C18-H-49KPDMS	----	----	4.38	0.08	0.26	0.53	0.53
1/1C14/C18-H-2KPDMS	3.5	----	----	0.15	0.17	0.53	0.53
1/1C14/C18-H-18KPDMS	----	3.5	----	0.15	0.17	0.53	0.53
1/1C14/C18-H-49KPDMS	----	----	4.38	0.15	0.17	0.53	0.53
3/1C14/C18-H-2KPDMS	3.5	----	----	0.23	0.09	0.53	0.53
3/1C14/C18-H-18KPDMS	----	3.5	----	0.23	0.09	0.53	0.53
3/1C14/C18-H-49KPDMS	----	----	4.38	0.23	0.09	0.53	0.53

Bacterial Biofilm Retention Assay

A description of the bacterial biofilm retention assay has been previously described in detail.[9,10] Coatings were preconditioned in a circulating deionized water tank until the coating leachates were determined to non-toxic to the bacteria. After that, the evaluation of bacterial biofilm retention on coating surfaces was carried out as follows: Coating plates were inoculated with a 1.0 mL suspension of bacterium in BGM ($\sim 10^7$cells/mL). The plates were then incubated statically in a 28°C incubator for 18 hours to facilitate bacterial attachment and subsequent colonization. The plates were then rinsed three times with 1.0 mL of deionized water to remove any planktonic or loosely attached biofilm. The biofilm retained on each coating surface after rinsing was then stained with crystal violet. Once dry, the crystal violet dye was extracted from the biofilm with the addition of 0.5 mL of glacial acetic acid and the resulting eluate was measured for absorbance at 600 nm. The absorbance values obtained were directly proportional to the amount of biofilm retained on the coating surface. Each data point was reported as the mean absorbance value of three replicate wells as a ratio to the silicone reference coating. Error bars represent one standard deviation of the absorbance ratio. Coated catheter sections were also evaluated against different microorganisms by following the same procedure.

RESULTS AND DISCUSSION

Polysiloxane coatings containing tethered QAS moieties were prepared by solution blending silanol-terminated polydimethylsiloxanes (PDMS), QAS-functional trimethoxysilanes, MeTAS, and TBAF catalyst solution and depositing coating solutions into modified microtiter array plates. As illustrated in Figure 1, crosslinking and tethering of QAS moieties occurs via a mixture of condensation reactions involving Si-OH, acetoxysilane, and methoxysilane groups.

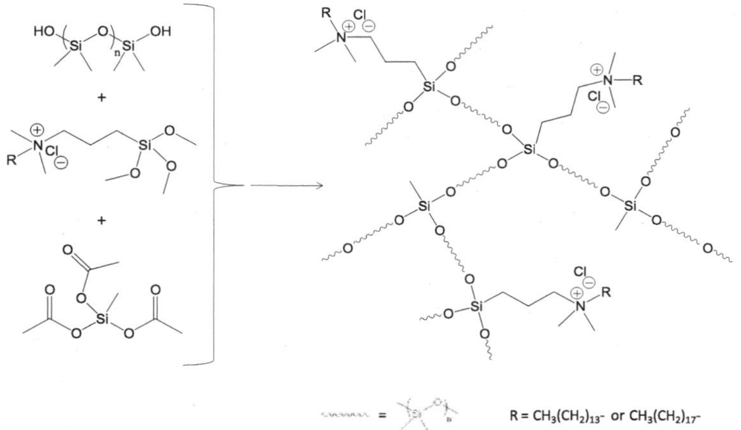

R = $CH_3(CH_2)_{13}$- or $CH_3(CH_2)_{17}$-

Figure 1. The synthetic scheme describing the production of polysiloxane coatings containing tethered QAS moieties.

A library of 12 coatings was prepared based on variations in PDMS molecular weight, QAS composition, and QAS concentration. With regard to QAS composition, blends of C14-QAS and C18-QAS were used. The decision to use blends of C14-QAS and C18-QAS was based on the desire to generate broad spectrum antimicrobial activity and prior results that demonstrated specificity between antimicrobial activity and QAS alkyl chain length.[8] At the high level of QAS concentration (0.20 moles/Kg of PDMS), the ratio of C14-QAS to C18-QAS was varied from 1:3 to 3:1 on a molar basis. At the low level of QAS concentration (0.10 moles/Kg of PDMS), a 1:1 molar ratio of C14-QAS to C18-QAS was used. QAS chemical structure and an outline of the library design are shown in Figure 2.

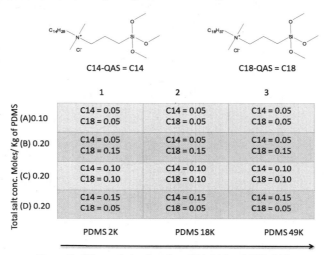

Figure 2. Library design based on C14-QAS and C18-QAS.

The antimicrobial activity of the coatings were evaluated toward the Gram-negative bacteria, *Pseudomonas aeruginosa* and *Escherichia coli*, and the Gram-positive bacterium, *Staphylococcus epidermidis*. The choice of these bacteria was based on their implication in urinary tract infections. The results of biofilm retention are shown in Figure 3.

Compared to the silicone reference coating, all the coatings showed moderate to high reduction in biofilm retention toward *P. aeruginosa* (Figure 3a). The highest reduction in biofilm retention (96% reduction compared to DC3140) was obtained with coating 1/3C14/C18-H-49KPDMS (coating B3 in the library). Coatings possessing the highest C14-QAS concentration were not active toward *E. coli* (Figure 3b). However, the rest of the coatings with equal or higher concentration of C18-QAS had moderate to high activity toward *E. coli*. The highest activity toward *E. coli* was obtained with coating 1/3C14/C18-H-49KPDMS (87% reduction compared to DC3140). With respect to the Gram-positive bacterium, *S. epidermidis*, only coating 1/3C14/C18-H-49KPDMS displayed high antimicrobial activity.

Evaluation of the antimicrobial activity of these twelve coatings toward the three different microorganisms resulted in the identification of a single coating composition

(1/3C14/C18-H-49KPDMS) that possessed broad spectrum antimicrobial activity. This coating was used to coat sections of urinary catheters to determine the effect of the coating on antimicrobial activity.

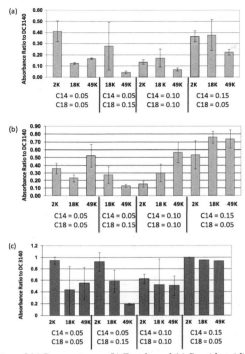

Figure 3. Evaluation of (a) P. aeruginosa, (b) E. coli, and (c) S. epidermidis biofilm retention.

Antimicrobial activity of coated catheters was evaluated using *P. aeruginosa, E. coli, S. epidermidis,* and *Candida albicans* (fungus implicated in catheter related infections). The images of crystal violet stained coated catheter sections after testing toward *E. coli* is shown in Figure 4. The results of biofilm retention measurements of the QAS-coated catheter sections compared to uncoated catheter sections are shown in Figure 5. The results displayed in Figure 5 show that coating the sections of urinary catheters with the optimized coating, 1/3C14/C18-H-49KPDMS, resulted in significant reduction in biofilm retention.

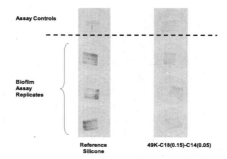

Figure 4. Images of crystal violet stained QAS-coated catheter sections after inoculation with E. coli.

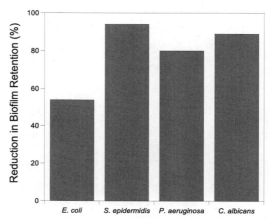

Figure 5. The reduction in microbial biofilm retention on QAS-coated catheter sections. The numerical values represent relative reduction in biofilm retention compared to the uncoated catheter.

CONCLUSIONS

A library of twelve QAS-tethered siloxane coatings based on combinations of C14-QAS and C18-QAS were prepared and the antimicrobial activity of these coatings evaluated toward *P. aeruginosa, E. coli,* and *S. epidermidis.* The coating based on 49K PDMS, 0.05 moles/Kg C14-QAS, and 0.15 moles/Kg C18-QAS was determined to be the most effective and, thus, was used to coat urinary catheters. The results showed that the percent reduction in biofilm formation obtained with coated catheters relative to uncoated coated catheters was 54% toward *E. coli,* 94% toward *S. epidermidis*, 80% toward *P. aeruginosa*, and 89% toward *C. albicans.*

ACKNOWLEDGMENTS

The authors acknowledge financial support from the Office of Naval Research under grants N00014-05-1-0822 and N00014-06-1-0952.

REFERENCES

[1] G. Sauvet, S. Dupond, K. Kazmierski, and J. Chojnowski, Biocidal Polymers Active by Contact. V. Synthesis of Polysiloxanes with Biocidal Activity, *J. Appl. Polym. Sci.*, Vol 75, 2000, p 1005-1012.

[2] J. Hazziza-Laskar, G. Helary, and G. Sauvet, Biocidal Polymers Active by Contact. IV. Polyurethanes based on Polysiloxanes with Pendant Primary Alcohols and Quaternary Ammonium Groups, *J. Appl. Polym. Sci.*, Vol. 58(1), 1995, p 77-84.

[3] N. Nurdin, G. Helary, and G. Sauvet, Biocidal Polymers Active by Contact. II. Biological Evalution of Polyurethane Coatings with Pendant Quaternary Ammonium Salts, *J. Appl. Polym. Sci.*, Vol 50, 1993, p 663-670.

[4] A. Russell, The Mechanism of Action of some Antibacterial Agents, *Prog. Med. Chem.*, Vol 6, 1969, p 135-199.

[5] B. Chisholm, R. Potyrailo, J. Cawse, R. Shaffer, M. Brennan, C. Molaison, D. Whisenhunt, B. Flanagan, D. Olson, J. Akhave, D. Saunders, A. Mehrabi, and M. Licon, The Development of Combinatorial Chemistry Methods for Coating Development. I: Overview of the Experimental Factory, *Prog. Org. Coat.,* Vol 45, 2002, p 313-321.

[6] P. Majumdar, S. Stafslien, J. Daniels, and D. Webster, High-Throughput Combinatorial Characterization of Thermosetting Siloxane-Urethane Coatings Having Spontaneously Formed Microtopographical Surfaces, *JCT-Research*, Vol 4(2), 2007, p 131-138.

[7] D. Webster, Radical Change in Research and Development: The Shift from Conventional Methods to High-Throughput Methods, *JCT CoatingsTech*, Vol 2, 2005, p 24-29.

[8] P. Majumdar, E. Lee, N. Patel, K. Ward, S. Stafslien, J. Daniels, B. Chisholm, P. Boudjouk, M. Callow, J. Callow, and S. Thompson, Combinatorial Materials Research Applied to the Development of New Surface Coatings IX: An Investigation of Novel Antifouling/Fouling-Release Coatings Containing Quaternary Ammonium Salt Groups, *Biofouling*, Vol 24(3), 2008, p 185-200.

[9] S. Stafslien, J. Daniels, B. Chisholm, and D. Christianson, Combinatorial Materials Research Applied to the Development of New Surface Coatings III: Utilization of a High-Throughput Multiwell Plate Screening Method to Rapidly Assess Bacterial Biofilm Retention on Antifouling Surfaces, *Biofouling*, Vol 23(1/2), 2007, p 37-44.

[10] S. Stafslien, J. Daniels, B. Mayo, D. Christianson, B. Chisholm, A. Ekin, D. Webster, and G. Swain, Combinatorial Materials Research Applied to the Development of New Surface Coatings IV: A High-Throughput Bacterial Biofilm Retention and Retraction Assay for Screening Fouling-Release Performance of Coatings, *Biofouling*, Vol 23(1), 2007, p 45-54.

HIGH-THROUGHPUT MICROBIAL BIOFILM ASSAY FOR THE RAPID DISCOVERY OF ANTIMICROBIAL COATINGS AND MATERIALS FOR BIOMEDICAL APPLICATIONS

S. Stafslien, B. Chisholm, P. Majumdar, J. Bahr, J. Daniels

Center for Nanoscale Science and Engineering, North Dakota State University, Fargo, ND, USA

ABSTRACT

A significant effort has been initiated at North Dakota State University to develop new coating technologies that effectively minimize or prevent the un-wanted microbial colonization of implanted medical devices. The primary focus of this research program has been aimed at the development of contact active or non-leaching coating systems using a combinatorial approach to dramatically accelerate the discovery process. A high-throughput microbial biofilm assay has been developed and leveraged as a primary screening tool to rapidly identify coating compositions that exhibit promising antimicrobial properties. A library of polysiloxane coatings containing bound ammonium salt moieties was generated using the combinatorial workflow and evaluated with the high-throughput microbial biofilm assays. Several compositions exhibited a substantial reduction in *Escherichia coli* and *Staphylococcus epidermidis* biofilm retention, without the presence of leachate toxicity, and were identified as coating candidates that could potentially be utilized to combat medical device-related infections.

INTRODUCTION

The authors have been heavily focused on the development and application of combinatorial, high-throughput methods to polymer and surface coating research and development [1]. The current workflow is being utilized to develop new environmentally-friendly coatings for ship hulls that deter settlement of marine organisms without leaching toxic compounds into the aquatic environment [2]. A general material approach has involved the development of moisture-curable, hybrid siloxane coatings containing either bound ammonium salt groups or tethered triclosan moieties as non-leachable biocide moieties [3, 4]. This same approach could potentially be utilized to impart antimicrobial activity to invasive biomedical devices, such as indwelling urinary catheters, vascular stints, and endotracheal tubes. These coating systems may be an attractive alternative to active-release technologies as they have been hypothesized to be less prone to antimicrobial resistance and have the potential to maintain long-term or permanent antimicrobial activity [5].

The authors have previously reported on the development of a multi-well plate, microbial biofilm screening assay used to quickly characterize the antifouling properties of marine coatings [6-9]. The multi-well plate assay has been recently modified to utilize biomedically relevant microorganisms for rapidly determining the antimicrobial properties of coatings for biomedical applications (Figure 1). This document reports on the utility of the multi-well plate assay to effectively screen a library of siloxane coatings containing bound quaternary ammonium salt groups with both a Gram-positive (*Staphylococcus epidermidis*) and Gram-negative bacterium (*Escherichia coli*) known to be associated with device related infections.

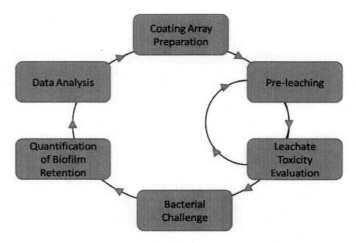

Figure 1. Diagram of the multiwell plate, microbial biofilm screening assay.

METHODS AND MATERIALS

Coating Preparation

Quaternary ammonium salts containing trimethoxysilane groups (QAS) were reacted with silanol-terminated polydimethylsiloxane (PDMS) polymers and methyltriacetoxysilane using tetrabutylammonium fluoride as a catalyst to form coatings containing QAS groups covalently bound (tethered) to a crosslinked PDMS matrix. The experimental design, shown in Figure 2, involved the generation of coating libraries that possessed variations in silanol-terminated PDMS molecular weight, QAS composition, and QAS concentration. The three different silanol-terminated PDMS molecular weights used for the study were 2,000, 18,000, and 49,000 g/mole. Five different QASs were used at two different concentrations. The QASs varied with respect to alkyl chain length from C10 to C18, as shown in Figure 2. The two QAS concentrations utilized were 0.10 moles and 0.20 moles per Kg of silanol-terminated PDMS. For each coating, the concentration of methyltriacetoxysilane was kept constant at 15 wt% of the silanol-terminated PDMS content.

An automated coating formulation system manufactured by Symyx Discovery Tools, Inc. was used to prepare coating solutions. Coatings were dispensed into 8 ml vials using a liquid handling robot possessing interchangeable tips and mixed with a magnetic stir bar in each vial. 0.25 mL of each coating solution was deposited into one column of a 24 well polystyrene plate modified with epoxy primed aluminum discs [6]. Each coating plate also contained a silicone reference coating (35% by weight suspension of a silicone elastomer (SE) control in methylisobutylketone) which was used to compare coating performance among the plates. Coatings were cured for 24 hours at room temperature, followed by an additional 24 hour treatment at 50° C.

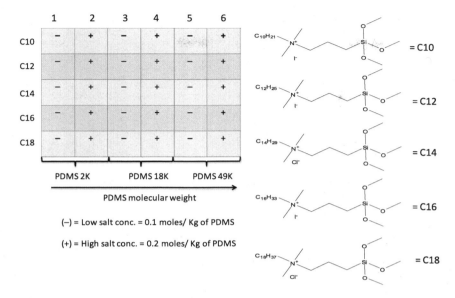

Figure 2. Design of PDMS-QAS coatings library (left). Chemical structure of QAS (right).

Evaluation of Leachate Toxicity

Coating array plates were pre-leached in a recirculating deionized water tank for two weeks to remove any leachable residues. 1 ml of growth medium was added to each well and placed on an orbital shaker overnight at ambient conditions. The coating extracts were inoculated with 50 µl of *E. coli* or *S. epdidermidis* (~10^7 cells.ml^{-1}) and 200 µl aliquots were transferred in triplicate to a 96-well plate. Plates were incubated for 24 hr at 28°C, rinsed three times with deionized water, and stained with crystal violet. Crystal violet stained biofilms were extracted with acetic acid and 150 µl aliquots of the resulting eluate were transferred in triplicate to a new 96-well plate and measured for absorbance at 600 nm. A growth positive (bacterium in fresh growth medium) and growth negative (bacterium in fresh growth medium spiked with 1µg.ml^{-1} of triclosan) control were included in each leachate toxicity plate.

Bacterial Biofilm Retention Assay

After completion of the leachate toxicity assay, coating array plates were evaluated for their antimicrobial activity using a multiwell plate, bacterial biofilm screening assay described previously [6, 7]. Briefly, coating array plates were inoculated with a 1.0 mL suspension of *E. coli* or *S. epidermidis* in growth medium (~10^7 cells.ml^{-1}). The plates were then incubated statically in a 28° C incubator for 18 hours to facilitate bacterial attachment and subsequent colonization. The plates were then rinsed three times with 1.0 mL of deionized water to remove any planktonic or loosely attached biofilm. The biofilm retained on each coating surface after rinsing was then stained with crystal violet dye. Once dry, the crystal violet dye was extracted from the biofilm with the addition of 0.5 mL of acetic acid and 150 µl aliquots of the resulting eluate were transferred in triplicate to a new 96-well plate and measured for absorbance at 600 nm. The absorbance values were directly proportional to the amount of biofilm retained on the coating surface. Each data point represented the mean absorbance value of three replicate samples and was reported as a ratio to the SE control coating contained in each array plate. Error bars represent one standard deviation of the mean.

RESULTS AND DISCUSSION

When evaluating the antimicrobial properties of coatings designed to function by contact, rather than release of a biocidal agent, it is important to demonstrate that no toxic components leach from the coating matrix when immersed in an aqueous medium. Failure to carry out this analysis can lead to false positives during the screening process. Figure 3 shows a representative image of the *S. epidermidis* and *E. coli* leachate toxicity analysis carried out on a subset of the PDMS coatings containing bound QAS groups (PDMS-QAS). All of the PDMS-QAS coatings showed comparable growth in the coating leachates to that of the growth positive controls for both *E. coli* and *S. epidermidis*. These results confirm that no toxic components leached into the overlying growth medium during the overnight extraction and any reduction in biofilm retention observed on the coating surfaces can be attributed to a surface associated phenomenon.

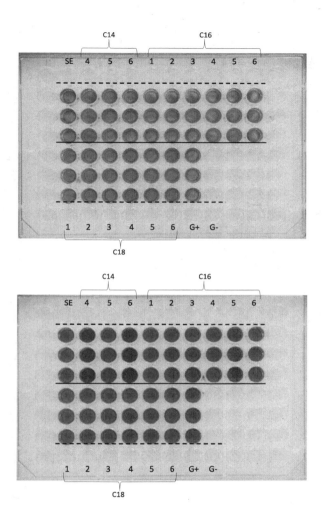

Figure 3. Image of E. coli (top) and S. epidermidis (bottom) leachate toxicity evaluations in a 96-well plate. Three replicate sample wells (between dashed and solid line) and one assay control (between edge of plate and dashed line) for each coating leachate. G + = positive growth control, G- = negative growth control.

Figure 4 shows some representative images of the PDMS-QAS coating array plates after bacterial incubation and staining with crystal violet. Visual inspection clearly demonstrates that the coatings with the bound C16 QAS showed little or no activity with regards to reducing the biofilm retention on the coating surfaces. However, coatings containing the C18 QAS exhibited activity for both bacteria.

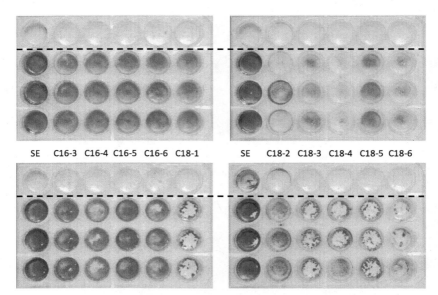

Figure 4. Images of crystal violet stained E. coli (top) and S. epidermidis (bottom) biofilms retained on the PDMS-QAS coatings cast in multi-well plates. The top row of the plate (wells above the dashed line) served as an assay control for each PDMS-QAS coating (no bacterial biofilm).

The crystal violet extraction data for the PDMS-QAS coatings are shown in Figure 5. As noted from the images in Figure 4 and the data in Figure 5, the coatings based on the C18 QAS showed a higher degree of activity than those based on the C16 QAS. However, it is also evident that the C14 QAS provided some degree of activity with respect to *S. epidermidis*, but no activity was observed for *E. coli*. When examining the effect of QAS concentration in the most active coatings (i.e., those containing the C18 QAS) the higher QAS concentration, 0.20 moles/Kg of PDMS, showed an increased reduction in biofilm retention. The molecular weight of the silanol-terminated PDMS appeared to have no discernable impact on the antimicrobial activity of the coatings.

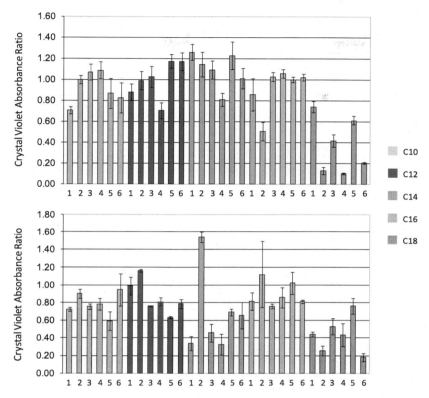

Figure 5. Crystal violet extraction data for E. coli (top) and S. epidermidis (bottom) biofilms retained on the coating surfaces. X-axis indicates unique coating compositions for each QAS group (described in Figure 2).

CONCLUSIONS

A high-throughput microbial biofilm screening assay was developed to rapidly analyze the antimicrobial properties of coatings developed using a combinatorial approach. Libraries of coating arrays are cast in 24-well plates and analyzed for their propensity to release toxic components and their ability to resist microbial biofilm retention. A series of PDMS-QAS coatings were evaluated with the multi-well pate screening assay and the results obtained identified coatings containing 0.2 moles C18 QAS/Kg of PDMS as having good broad spectrum activity towards *E. coli* and *S. epidermidis* without the presence of leachate toxicity. As a result, the PDMS-QAS coatings may offer a highly effective solution to combat device-related infections.

ACKNOWLEDGMENTS

The authors would like to thank the Office of Naval Research for supporting this research under Grants N00014-05-1-0822 and N00014-06-1-0952.

REFERENCES

[1] http://www.ndsu.edu/cnse/facilities_equipment/bioactive.html

[2] D. Webster, B. Chisholm and S. Stafslien, Combinatorial Approaches for Novel Coatings Design: A Mini-Review, *Biofouling*, 23, 2007, p 179-192.

[3] P. Majumdar, E. Lee, N. Patel, K. Ward, S. Stafslien, J. Daniels, B. Chisholm, P. Boudjouk, M. Callow, J. Callow, S. Thompson, Combinatorial Materials Research Applied to the Development of New Surface Coatings IX: An Investigation of Novel Anti-Fouling/Fouling-Release Coatings Containing Quaternary Ammonium Salt Groups, *Biofouling*, 24(3), 2008, p 185-200.

[4] J. Thomas, S. Choi, R. Fjeldheim and P. Boudjouk, Silicones Containing Pendant Biocides for Antifouling Coatings, *Biofouling*, 20, 2004, p 227-236.

[5] K. Lewis and A. Klibanov, Surpassing Nature: Rational Design of Sterile-Surface Materials, *TRENDS in Biotechnology*, 23, 2005, p 343-348.

[6] S. Stafslien, J. Bahr, J. Feser, J. Weisz, B. Chisholm, T. Ready and P. Boudjouk, Combinatorial Materials Research Applied to the Development of New Surface Coatings I: A Multiwell Plate Screening Method for the High-Throughput Assessment of Bacterial Biofilm Retention on Surfaces, *J. Comb. Chem.*, 8, 2006, p 156-162.

[7] S. Stafslien, J. Daniels, B. Chisholm and D. Christainson, Combinatorial Materials Research Applied to the Development of New Surface Coatings III: Utilization of a High-Throughput Multiwell Plate Screening Method to Rapidly Assess Bacterial Biofilm Retention on Antifouling Surfaces, *Biofouling*, 23, 2007, p 37-44.

[8] S. Stafslien, J. Daniels, B. Mayo, D. Christianson, B. Chisholm, A. Ekin, D. Webster and G. Swain, Combinatorial Materials Research Applied to the Development of New Surface Coatings IV: A High-Throughput Bacterial Biofilm Retention and Retraction Assay for Screening Fouling-Release Performance of Coatings, *Biofouling*, 23, 2007, p 45-54.

[9] E. Ribeiro, S.J. Stafslien, F. Cassé, J.A. Callow, M.E. Callow, R.J. Pieper, J.W. Daniels, J.A. Bahr and D.C. Webster, Automated image-based method for laboratory screening of coating libraries for adhesion of algae and bacterial biofilms, *J. Combi. Chem.*, 10(4), 2008, p 586-594.

CHEMICAL - HYDROTHERMAL COMBINED SYNTHESIS OF BIOACTIVE TiO$_2$ AND CaTiO$_3$ FILMS ON Ti SURFACES

M. Ueda[a,*], M. Ikeda[a] and M. Ogawa[b]

[a] Faculty of Chemistry, Materials and Bioengineering, Kansai University,

3-3-35 Yamate-cho, Suita, Osaka 564-8680, Japan

[b] R&D laboratory, Daio Steel Co., Ltd,

30 Daido-cho 2-chome, Minami-ku, Nagoya, Aichi 457-8545, Japan

ABSTRACT

The chemical-hydrothermal combined synthesis of TiO$_2$ and CaTiO$_3$ films on pure Ti substrates was examined with a focus on crystallinity and surface morphology of the films. Pure Ti disks were chemically treated with H$_2$O$_2$/HNO$_3$ solutions at 353 K for 20 min in order to introduce a TiO$_2$ layer with low crystallinity on the surface. The samples were then hydrothermally treated in an autoclave at 453 K for 12 h or 24 h. Anatase-type TiO$_2$ and perovskite-type CaTiO$_3$ films with high crystallinity were obtained upon treatment with distilled water and an aqueous solution of Ca(OH)$_2$, respectively. Cracks in the TiO$_2$ precursor films disappeared after hydrothermal treatment. Uniform and crack-free films could be obtained by the present process. In addition, in vitro formation of hydroxyapatite (HAp) on the films was investigated. Samples were immersed in SBF (Simulated Body Fluid), adjusted to 310 K. A light HAp precipitate could be observed on non-surface modified Ti after 2 days of immersion. In contrast, precipitates formed after only 2 days on the oxide films. The present surface modifications were confirmed to drastically promote the deposition of HAp on the surface. The surfaces also show excellent adhesion of osteoblast-like MC3T3E1 cell.

INTRODUCTION

Titanium and its alloys have been widely used as biomaterials for hard tissue substitutes because of their excellent mechanical properties and biocompatibility. However, the osteointegration of these metallic materials is less than that of bioactive ceramics. Therefore, various surface modification techniques have been developed to improve the osteointegration of these materials [1-10]. The simplest way is to synthesis the bioactive ceramic films on titanium or its alloys with required mechanical properties.

In general, ceramic films are prepared on substrates by methods such as plasma spraying, sputtering, and sol-gel [11-17]. Although such conventional methods are well developed, several problems remain with these methods. For example, plasma spraying and sputtering require large-scale and expensive equipment, and the sol-gel method requires strict control of the atmosphere in order to suppress hydrolysis of the starting solutions. Therefore, we have sought to determine new methods of preparation for oxide films. In previous paper, the preparations of TiO$_2$ and SrTiO$_3$ films by oxidation of Ti plates were reported [16,17]. Although this is a very simple route to produce these films, heat treatment at high temperature is necessary to induce oxidation of the substrates.

Recently, surface modifications by chemical reactions have been very popular in the design of biomaterials, for example, chemical treatment with NaOH or H_2O_2 solutions [1-3]. Moreover, one common chemical modification, NaOH treatment, has been clinically employed. Such a wet process is suitable for surface modification of substrates with complex shapes and/or large surface areas. However, the synthesis of ceramic films generally requires heat treatment at high temperature after the chemical treatments.

The hydrothermal technique is a wet chemical process that has been widely utilized for the preparation of nanocrystalline oxide materials such as $BaTiO_3$, ZrO_2 and TiO_2 [18-21]. Hydrothermal synthesis is carried out at a temperature between the boiling point and the critical point (647 K, 22.1MPa) of water. This process has also been employed in surface modification of biomaterials [9,10]. In particular, a number of researchers have attempted to produce $CaTiO_3$ films on Ti substrates using hydrothermal processes. $CaTiO_3$ is a favorable oxide from the viewpoint of inducing bone on the materials; calcium ions seem to enhance the precipitation of hydroxyapatite (HAp) $Ca_{10}(PO_4)_6(OH)_2$ on the surface. However, single-phase films of $CaTiO_3$ have not been accomplished on Ti substrates by a simple hydrothermal process in the absence of an external energy supply.

Recently, we have investigated the chemical-hydrothermal combined synthesis of oxide films by dissolving and precipitating titanium on the surface of titanium and its alloy substrates with a focus on crystallinity, surface morphology and thickness of the films [22].

In the current work, we selected TiO_2 and $CaTiO_3$ as bioactive coating materials on pure Ti substrates. The purpose of the present work was to synthesize TiO_2 and $CaTiO_3$ films on pure Ti substrates by chemical-hydrothermal combined treatment with distilled water and aqueous solutions of $Ca(OH)_2$, respectively. In addition, formation of HAp and adhesion of osteoblast-like MC3T3E1 cells on the modified Ti surfaces were also investigated

MATERIALS AND METHODS

Pure Ti disks (ϕ 6 mm in diameter, 2 mm in thickness) were cut from a rod and mechanically polished with #400-1500 emery papers and 0.3 μm alumina paste. The disks were then chemically treated with 5 M H_2O_2/0.1 M HNO_3 (2.5 mL for each sample) at 353 K for 20 min. The samples were placed in a Teflon-lined autoclave with an internal volume of 50 mL (8 disks per batch), which was then filled with distilled water or 20 mM $Ca(OH)_2$ aqueous solution up to 50 % volume. The reactor was maintained at 453 K for 12 h or 24 h, and then allowed to cool down naturally. The samples were rinsed in distilled water and dried at 353 K for 30 min. Some samples were heat treated at 673 K for 1 h in air, and then allowed to cool down in a furnace.

In the current work, the following notations are used for the sake of simplicity: "Chem-", "Hyd-" and "HT-" represent chemical, hydrothermal and heat treatments, respectively. It is also important to note when "Chem-" is appropriately omitted, because chemical treatment is always carried out before hydrothermal treatment.

Hanks' solution with a nearly equal ion concentration to that of human blood plasma (Na^+ 142.0, K^+ 5.8, Mg^+ 0.9, Ca^+ 1.3, Cl^- 145.6, HCO_3^- 4.2, HPO_4^{2-} 0.8, SO_4^{2-} 0.4 mM) was employed as SBF (Simulated Body Fluid). Samples were immersed in the SBF (5 mL per disk), adjusted to 310 K. After soaking for different periods of time, up to 20 days, the samples were washed with distilled water and then dried at 323 K for 3 h. The SBF was changed every 2 days.

Osteoblast-like MC3T3E1 cells were seeded at 2×10^5 cells/mL on samples in 96-well micro-plates. These cells were maintained in 150 μL of DMEM (Dulbecco-modified Eagle's medium) containing 10% FBS (fetal bovine serum) and antibiotics penicillin-streptomycin. The cells on samples were incubated at 310 K for 2 h in a controlled humidified 5% CO_2 atmosphere. After suction of supernatant fluid to remove floating cells, cell nuclei are stained by 0.04% crystal violet. Samples were rinsed three times with PBS (phosphate buffered saline), and the cells were separated from the substrates using DMSO (dimethyl sulfoxide). The separated cells were suspended in distilled water and absorbance was measured at wavelength of 450 nm. The adhesion of cells was assessed by Student's t-test.

Low angle X-ray diffraction (XRD, Cu Kα radiation) analysis was performed using a Rigaku RINT2500 at an incident angle of 1 degree. Scanning electron microscopy (SEM) images were taken using a JEOL JSM-6500F.

RESULTS AND DISCUSSION

Synthesis of oxide films on Ti surface

As previously noted, there are several routes to prepare oxide films, such as sputtering, sol-gel and anodic oxidation [11-17]. In the current work, preparation of bioactive TiO_2 and $CaTiO_3$ films with high crystallinity was attempted at low temperature on pure Ti substrates using chemical-hydrothermal combined treatments.

Titanium is known to dissolve in H_2O_2 solution according to the following [23]:

$$Ti + 3H_2O_2 \rightarrow [Ti(OH)_3O_2]^- + H_2O + H^+ \qquad (1)$$

This forward reaction suggests that a reverse reaction can be induced by pH control. The dissolved titanium is thought to precipitate as titanium oxide under low pH conditions. A titanium oxide film on the Ti surface was previously found to form upon chemical treatment with an aqueous solution of H_2O_2. However, the formed film was not homogenous. Wang et al. reported that a TiO_2 gel layer could be obtained on Ti substrates by chemical treatment with a H_2O_2/HCl solution [3]. Further, addition of HCl to H_2O_2 was reported to catalyze the reaction and generate a uniform gel over the entire surface. This finding was in good agreement with the results of the current work. In our previous work [22], we compared the formation of the TiO$_2$ gel layer by chemical treatment with 9 M H$_2$O$_2$/0.1 M HCl and 9 M H$_2$O$_2$/0.1 M HNO$_3$ solution in pure Ti and Ti-Nb alloys. HNO$_3$ is superior to HCl in its ability to break passive films on the surface of Ti alloys. Although lifting of the TiO$_2$ films from Ti-Nb alloy substrates was observed in the H$_2$O$_2$/HCl treatment, a uniform film without cracks could be obtained by the H$_2$O$_2$/HNO$_3$ treatment. A 5 M H$_2$O$_2$/0.1 M HNO$_3$ solution was employed in the chemical treatments in this work. The concentration of H$_2$O$_2$ was diluted to 5 M from 9 M in order to slightly reduce the thickness of the Chem-TiO$_2$ films.

Figure 1 shows surface morphologies of the Ti substrates after chemical treatment with the H$_2$O$_2$/ HNO$_3$ solution at 353 K for 20 min. Uniform films could be formed, although a small amount of discontinuous micro-cracks were observed, as shown in Fig. 1(a). The surface of the Ti substrate was not uniformly covered with the film; the substrate was partially exposed for treatments shorter than 10 min. On the other hand, chemical treatment longer than 40 min resulted in the formation of continuous cracks in the TiO$_2$ film. The surfaces of the films show a sponge-like morphology with large surface area [Fig. 1(b)] irrespective of the chemical treatment time. The morphology of the films was quite similar to those obtained from NaOH treatment, although the present morphology is more refined. Moreover, no lifting of the films was observed during the treatments.

Following the chemical treatment, hydrothermal treatments with distilled water or aqueous Ca(OH)$_2$ were carried out. The resultant films prepared from the chemical treatment for 20 min were employed as precursors. The thickness of the TiO$_2$ precursor film was determined to be about 400 nm from a cross-sectional SEM image. Figure 2 shows XRD profiles of the pure Ti substrate and the resultant films prepared by chemical and hydrothermal treatments. No reflections from the resultant film prepared by chemical treatment were observed, as shown in Fig. 2 (b). However, XRD peaks from the Ti substrate become small compared with non-surface modified Ti [Fig. 2(a)]. Although the XRD profiles are not shown here, the broad peaks corresponding to anatase-type TiO$_2$ were confirmed to appear after the chemical treatment for longer than 60 min. The substrate was determined to be covered with anatase-type TiO$_2$ with low crystallinity. On the other hand, anatase-type TiO$_2$ and

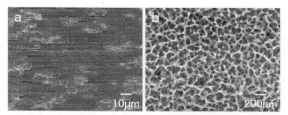

Figure 1. SEM images of surface product on pure Ti substrates after chemical treatment with H$_2$O$_2$/HNO$_3$ aqueous solution for 20 min (a). (b) is a magnified image of (a).

perovskite-type CaTiO$_3$ films with high crystallinity were confirmed to form upon the hydrothermal treatments with distilled water and aqueous Ca(OH)$_2$, respectively (**Fig. 2 (c) and (d)**). In the CaTiO$_3$ film, a small amount of unreacted Ca(OH)$_2$ was confirmed to remain. Typically, synthesis of such oxide films with high crystallinity requires high temperature heating. By the present route, the crystallized oxide films could be easily obtained at low temperatures.

    Figure 3 shows the morphology of the TiO$_2$ and CaTiO$_3$ films. After the hydrothermal treatment with distilled water, the sponge-like morphology of the TiO$_2$ gel (**Fig. 1**) was maintained. This morphology consisted of very fine cube crystals with an average size of approximately 10 nm as

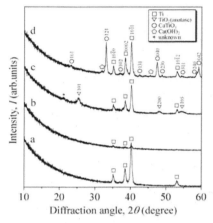

Figure 2. X-ray diffraction patterns from (a) pure Ti substrate, (b) TiO$_2$ prepared by chemical treatment and product films prepared by hydrothermal treatment with (c) distilled water and (d) aqueous Ca(OH)$_2$, respectively.

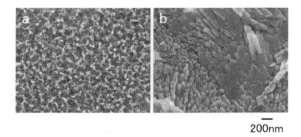

200nm

Figure 3. SEM images of hydrothermally synthesized (a) TiO$_2$ and (b) CaTiO$_3$ films.

shown in **Fig. 3(a)**. In contrast, the morphology of the surface was drastically changed after hydrothermal treatment with Ca(OH)$_2$ solution. Fine rectangular prism-shaped crystals are formed in the Hyd-CaTiO$_3$ films, as shown in **Fig. 3(b)**. The average length along the edge of the square faces of the CaTiO$_3$ crystals was determined to be approximately 50 nm. In general, titanium oxides are known to dissolve in alkaline solution. The dissolution of titanium oxides and precipitation of CaTiO$_3$ was strongly activated by hydrothermal treatment in aqueous alkaline solution of Ca(OH)$_2$. In the present process, the following points should be noted: "the size distribution is very narrow and the micro-cracks in the TiO$_2$ precursor film disappear after the hydrothermal treatments". Thus, uniform and crack-free films were obtained upon combined chemical-hydrothermal treatment. Although the images are not shown here, the morphologies were not changed before/after the heat treatment at 673 K.

In order to estimate the change in the crystallinity of the films, values of full width half-maximum (FWHM) intensity in the 200 peak of TiO$_2$ and the 040 peak of CaTiO$_3$ were measured as shown in **Fig. 4**. The broken lines indicate the FWHMs obtained from the TiO$_2$ and CaTiO$_3$ films after heat treatment at 673 K. In general, the FWHM can be utilized as an indicator of crystallinity; a sharper peak (lower FWHM) indicates higher crystallinity. The crystallinity is drastically increased after the hydrothermal treatment in both oxides. However, the progress of crystallization is different for the two films. The FWHMs for peaks from CaTiO$_3$ is achieved in the equivalent level obtained by heat treatment. This observation suggests that the crystallinity has been sufficiently increased by hydrothermal treatment even for 12 h. The solvent employed in the hydrothermal treatment shows a

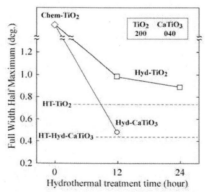

Figure 4. Values of full width at half-maximums intensity (FWHMs) of TiO$_2$ and CaTiO$_3$ in each state of chemical/hydrothermal/heat treatment. The values were calculated from 200 and040 reflections in XRD spectra for TiO$_2$ and CaTiO$_3$, respectively.

high pH, and the dissolution of TiO$_2$ and precipitation of CaTiO$_3$ are thought to occur actively compared with the synthesis of TiO$_2$ in distilled water.

HAp precipitation on the surface modified Ti

A simulated body fluid (SBF) nearly equal to that of human blood plasma was employed for in vitro bioactivity testing. Samples of surface-modified and non-modified Ti were immersed in the SBF, adjusted at 310 K. **Figure 5** shows the morphologies of the Hyd-TiO$_2$ **(a)** and Hyd- CaTiO$_3$ **(b)** after immersion in SBF for 2 days. The surfaces are covered with precipitates, which is identified to HAp by XRD analysis. Although the SEM image is not shown here, a very thin layer of HAp precipitates was observed on non-surface modified Ti after 6 days of immersion. The synthesized oxide films were confirmed to promote the deposition of HAp on the surface. In general, post heat treatment was always carried out after surface modification by a wet process, in order to introduce precipitation of HAp [1-3]. Interestingly, precipitation of HAp was observed on the surfaces not subjected to heat treatment. This result suggests that heat treatment is not necessary for the precipitation of HAp in the present surface modification process.

After immersion in SBF, various appearances of the formed HAp layer and the synthesized oxide films were observed and were classified into four different types; films that were fully or partially lifted from the substrates and films that were fully adhered with or without cracks. All samples were dried slowly at 323 K after immersion in SBF. Dehydration from the precipitated HAp or synthesized oxide films introduces internal stress at the interface between the HAp and the oxide films or the substrate. Therefore, the stability and adherence of films could be estimated from the appearance of the samples.

Precipitation of HAp and the appearance of the surface along with the immersion period in SBF for each sample are summarized in **Table I**, where "P" indicates the beginning of the HAp precipitation, "NC" represents the non-cracked samples, "MC" the discontinuous micro-cracks, "CC" the continuous cracks, "Lift (p)" the partial lifting and "Lift (f)" the full lifting of the films from the surface. The introduction of cracks in the HAp layer such as for the "MC" and "CC" samples, which are induced by the dehydration, is not a problem because the materials are employed in a wet environmental condition for practical use. Partial lifting was observed in the Chem- and HT-TiO$_2$ films after 6 days of soaking in SBF. In the Hyd-TiO$_2$ films, the HAp layer and synthesized films were fully lifted from the substrates. These observations suggest that the interface between the TiO$_2$ films and the substrates are less adhesive. The Hyd-TiO$_2$ film is quite uniform and crack-free as shown in **Fig. 3(b)**. However, crystallization may only occur very close to the surface. The TiO$_2$ film shows a duplex structure consisting of crystallized and non-crystallized (residual) TiO$_2$ layers. The duplex structure is therefore thought to increase the internal stress between the HAp and the Ti substrate. Thus, the precipitated HAp was fully lifted from the surface along with the synthesized TiO$_2$ film. In the fully

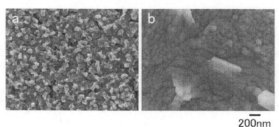

200nm

Figure 5. Surface morphologies of (a) Hyd-TiO$_2$ and (b) Hyd-CaTiO$_3$ films after immersion in SBF for 2 days.

Table I. Summary of HAp precipitations, introduction of cracks and exfoliation of films from substrates. "P" shows the beginning of the HAp precipitation, "NC" non cracks, "MC" discontinuous micro-cracks, "CC" continuous cracks, "Lift (p)" partial lifting and "Lift (f)" the full lifting of films, respectively.

Notation	Soaking period in SBF					
	2d	4d	6d	8d	14d	20d
Chem-TiO$_2$	P	CC	Lift (p)			→
HT-TiO$_2$	P, CC	CC	Lift (p)			→
Hyd-TiO$_2$	P	Lift (f)				→
HT-Hyd-TiO$_2$	P	NC	CC			→
Hyd-CaTiO$_3$	P	NC				→
HT-Hyd-CaTiO$_3$	P	NC	MC	→	CC	→
Ti			P	NC	MC	CC

crystallized HT-Hyd-TiO$_2$ films, the lifting of the films was not observed. This result supports the above conjecture. In the CaTiO$_3$ films, on the other hand, no lifting of films can be confirmed in the period of time in which the films are soaked in SBF. The CaTiO$_3$ surface is synthesized by hydrothermal treatment in a high pH solution. As a result, the crystallization is expected be complete. The values of FWHM for peaks in the XRD pattern from the CaTiO$_3$ film prepared by hydrothermal treatment were nearly equal to those obtained post heat treatment, as shown in **Fig. 4**. This suggests that Hyd-CaTiO$_3$ is a stable phase, because it is fully and uniformly crystallized. These results indicate that the CaTiO$_3$ prepared by the chemical-hydrothermal combined treatment is the most favorable candidate for the osteointegration from the viewpoint of the present in vitro test.

Adhesion of osteoblast-like MC3T3E1 cells on the modified Ti surfaces

The adhesion property in the incipient stage of osteoblast-like MC3T3E1 cells has been investigated. **Figure 6** shows the optical density in suspensions of adhered cells on the surface of Ti, Hyd-TiO$_2$ and Hyd-CaTiO$_3$. Statistically significant differences in the optical density were observed between the Hyd-TiO$_2$ and the Ti control. Hydrothermally synthesized TiO$_2$ greatly enhanced the

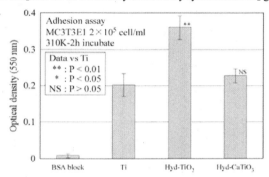

Figure 6. Adhesion assays of osteoblast-like MC3T3E1 cells on Ti (as control), Hyd-TiO$_2$ and Hyd-CaTiO$_3$ films.

adhesion of cells on the surface. On the other hand, Hyd-CaTiO$_3$ shows no significant difference from the Ti in the optical density. It can be said that the Hyd-CaTiO$_3$ film also shows good adhesion properties of MC3T3E1 cells since the Ti is generally known to possess an excellent biological affinity. Thus, the present surface modification for osteointegration enhanced or maintained the incipient adhesion ability of osteoblast-like cells compared with the Ti.

CONCLUSIONS

The chemical-hydrothermal combined modification of titanium surfaces, the formation of HAp and adhesion of osteoblast-like MC3T3E1 cells on the surfaces were investigated. The following conclusions were reached:

(1) Anatase-type TiO$_2$ films with low crystallinity were prepared on pure Ti substrates by chemical treatment with 5 M H$_2$O$_2$/0.1 M HNO$_3$ solution. The surface showed a sponge-like morphology with a large surface area.

(2) Anatase-type TiO$_2$ and perovskite-type CaTiO$_3$ films with high crystallinity could be synthesized by hydrothermal treatment from TiO$_2$ precursor films prepared by chemical treatment. The sponge-like morphology remained, and a duplex structure composed of crystallized and non-crystallized layers in the TiO$_2$ films was observed. In contrast, fully crystallized CaTiO$_3$ films could be obtained by hydrothermal treatment with a high pH solution. Fine rectangular prism-shaped crystals were uniformly observed in the CaTiO$_3$ films.

(3) A thin-layer of HAp was observed on non-surface modified Ti after 6 days immersion in SBF. On the surface-modified samples, in contrast, precipitates were observed after only 2 days. The present surface modification promotes the deposition of HAp on the surface of Ti.

(4) Hydrothermal treatment with Ca(OH)$_2$ solution using the TiO$_2$ precursor films is the most favorable candidate for the surface modification of pure Ti for improvement of osteointegration from the viewpoint of structural stability and ability to introduce HAp precipitate in SBF.

(5) The present surface modification for osteointegration enhanced or maintained the incipient adhesion ability of osteoblast-like cells compared with the Ti.

REFERENCES

[1]T. Kokubo, F. Miyaji, H.-M. Kim, and T. Nakamura, Spontaneous apatite formation on chemically treated titanium metals, *J. Am. Ceram. Soc.*, **79**, 1127-1129 (1996).

[2]T. Kokubo, H.-M. Kim, F. Miyaji, H. Takadama, and T. Miyazaki, Ceramic-metal and ceramic-polymer composites prepared by a biomimetic process, *Composites: Part A*, **30**, 405 -409 (1999).

[3]X.-X. Wang, K. Hayakawa, K. Tsuru, and A. Osaka, Bioactive titania gel layers formed by chemical treatment of Ti substrate with a H$_2$O$_2$/HCl solution, *Biomaterials*, **23**, 1353-1357 (2001).

[4]K. Asami, N. Ohtsu, K. Saito, and T. Hanawa, CaTiO$_3$ films sputter-deposited under simultaneous Ti-substrate, *Surf. Coat. Technol.*, **200**, 1005-1008 (2005).

[5]N. Ohtsu, K. Saito, K. Asami, and T. Hanawa, Characterization of CaTiO$_3$ thin film prepared by ion-beam assisted deposition, *Surf. Coat. Technol.*, **200**, 5455-5461 (2006).

[6]M. Okido, K. Kuroda, M. Ishikawa, R. Ichino, and O. Takai, Hydroxyapatite coating on titanium by means of thermal substrate method in aqueous solution, *Solid State Ionics*, **151**, 47-52 (2002).

[7]D. Stojanovic, B. Jokic, Dj. Veljovic, R. Petrovic, P.-S. Uskokovic, and Dj. Janackovic, Bioactive glass-apatite composite coating for titanium implant synthesized by electrophoretic deposition, *J. Eur. Ceram. Soc.*, **27**, 1595-1599 (2007).

[8]K. De Groot, R. Geesingk, C.-P.-A.-T. Klein, and P. Serekian, Plasma sprayed coatings of hydroxyapatite, *J. Biomed. Mater. Res.*, **21**, 1375-1381 (1987).

[9]K. Hamada, M. Kon, T. Hanawa, K. Yokoyama, Y. Miyamoto, and K. Asaoka, Hydrothermal modification of titanium surface in calcium solutions, *Biomaterials*, **23**, 2265-2272 (2002).

[10]J.-P. Wiff, V.-M. Fuenzalida, J.-L. Arias, and M.-S. Fernandez, Hydrothermal-electrochemical $CaTiO_3$ coatings as precursor of a biomimetic calcium phosphate layer, *Mater. Lett.*, **61**, 2739-2743 (2007).

[11]M. Yamaguchi, S. Kuriki, P.-K. Song, and Y. Shigesato, Thin film TiO_2 photocatalyst deposited by reactive magnetron sputtering, *Thin Solid Films*, **442**, 227-231 (2003).

[12]P. Zeman, and S. Takabayashi, Effect of total and oxygen partial pressures on structure of photocatalytic TiO_2 films sputtered on unheated substrate, *Surf. Coat. Technol.*, **153**, 93-99 (2002).

[13]M. Takahashi, K. Tsukigi, T. Uchino, and T. Yoko, Enhanced photocurrent in thin film TiO_2 electrodes prepared by sol–gel method, *Thin Solid Films*, **388**, 231-236 (2001).

[14]S. Zhang, Y.-F. Zhu, D.-E. Brodie, Photoconducting TiO_2 prepared by spray pyrolysis using $TiCl_4$, *Thin Solid Films*, **213**, 265-270 (1992).

[15]M. Okuya, N.-A. Prokudina, K. Mushika, and S. Kaneko, TiO_2 thin films synthesized by the spray pyrolysis deposition (SPD) technique, *J. Eur. Cera. Soc.*, **19**, 903-906 (1999).

[16]M. Ueda, and S. O.-Y.-Matsuo, Preparation of tabular TiO_2-$SrTiO_{3-\delta}$ composite for photocatalytic electrode, *Sci. Tech. Adv. Mater.*, **5**, 187-193 (2004).

[17]M. Ueda, D. Ohzaike, and S. O.-Y.-Matsuo, Low temperature synthesis of TiO_2/$SrTiO_3$ films on Ti substrate, *Materials Science Forum*, **512**, 217-222 (2006).

[18]W.-J. Dawson, Hydrothermal synthesis of advanced ceramic powders, *Am. Cera. Soc. Bull.*, **67**, 1673-1678 (1988).

[19]H. Xu, and L. Gao, Hydrothermal synthesis of high-purity $BaTiO_3$ powders: control of powder phase and size, sintering density, and dielectric properties, *Mater. Lett.*, **58**, 1582-1586 (2004).

[20]Y.-V. Kolen'ko, B.-R. Churagulov, M. Kunst, L. Mazerolles, and C. Colbeau-Justin, Photocatalytic properties of titania powders prepared by hydrothermal method, *Appl. Catal. B*, **54**, 51-58 (2004).

[21]M.-A. McCormick, and E.-B. Slamovich, Microstructure development and dielectric properties of hydrothermal $BaTiO_3$ thin films, *J. Eur. Cera. Soc.*, **23**, 2143-2152 (2003).

[22]M. Ueda, Y. Uchibayashi, S. O.-Y.-Matsuo, and T. Okura, Hydrothermal synthesis of anatase-type TiO_2 films on Ti and Ti-Nb substrates, *J. Alloys and Compounds*, **459**, 369-376 (2008).

[23]M. Kakihara, M. Tada, M. Shiro, V. Petrykin, M. Osada, Y. Nakamura, Structure and stability of water soluble $(NH_4)_8[Ti_4(C_6H_4O_7)_4(O_2)_4] \cdot 8H_2O$, *Inorg. Chem.*, **40**, 891-894 (2001).

SURFACE MODIFICATION OF HYDROXYAPATITE : A REVIEW

Otto C. Wilson, Jr.
The Catholic University of America
Department of Biomedical Engineering
BONE/CRAB Lab
620 Michigan Ave, NE
Washington, DC 20064

ABSTRACT

Adsorption interactions with hydroxyapatite (HAp) play an integral role in various processes that contribute to the birth, structural development, maintenance, and healing and remodeling of bone and dental tissue. The wide extent of its interactions in biological systems has contributed to its utilization as a substrate for various adsorption and surface complexation events. Various molecules including natural and synthetic polymers, inorganic polymers, surfactants, pharmaceutical agents, and protein based adsorbents have been used to modify the surface of HAp for applications ranging from improved colloid stability to enhanced bioactivity. An overview of HAp adsorption interactions will be presented in this talk with a focus on fascinating applications in the area of bone inspired nanocomposites.

INTRODUCTION

The surface modification of HAp has been a major research effort in the Biomimetics, Orthopedics, and Nanomaterials Education/Composite Research for Advanced Biomaterials (BONE/CRAB) Lab. The research vision of the BONE/CRAB Lab involves the following three focus areas: 1.) Develop bone inspired nanocomposite (BINs) and methodologies to enhance the healing and remodeling of bone at the tissue, cell, and subcellular level. 2.) Determine the key surface and interfacial characteristics of BINs that best predict/determine optimum hard tissue bioactivity and 3.) Understand the role of surface modification and adsorption phenomena in bone healing and remodeling. Our initial efforts in developing BINs focused on the synthesis, surface modification, and characterization of nanophase HAp. In particular, we focused on surface modification agents to improve the colloid stability of HAp. The colloid stability of hydroxyapatite (HAp) is a very important issue in a number of biological and technological processes such as biomineralization, protein purification, calcification, development of hard tissue implants, and gene therapy. Liu and Nancollas[1] determined that the colloid stability of HAp is determined by acid-base interactions and the inherent hydrophobicity of HAp particles in suspension contributes to the tendency for HAp particles to agglomerate extensively. Aggregation phenomena tend to be exacerbated under physiological conditions due to the high ionic strength and proximity of the isoelectric point (iep) of HAp to physiological pH. This problem is magnified even more when nanoscale particles are used.

A number of surface modification agents have been used to address the issue of HAp colloid stability in our research. We have shown that dodecyl alcohol can be used to improve the colloid stability of HAp in nonaqueous solvents[2]. This ability to disperse HAp in nonaqueous solvents is useful for incorporating HAp in polymers that are soluble in nonaqueous solvents. We demonstrated that HAp/polypropylene carbonate (PPC) nanophase composites could be synthesized by dispersing dodecyl alcohol modified HAp in acetone and subsequently dissolving

PPC in the suspension[3]. The BONE/CRAB Lab research team has also enhanced the colloidal stability of HAp using silica[4], polyethyleneimine (pei)[5], gum Arabic (GA)[6], and chitosan[7]. Silica turned out to be a unique multifunctional surface modification agent. In addition to improving the colloid stability of HAp, it also contributed improved acid stability[4], and the potential for improvements in bioactivity based the role of silica in bone biomineralization[8] and the unique properties of bioglass[9], and the ability to form stable and strong interfacial bonds with suitable polymer matrices using silane coupling agents. A number of researchers have attempted to use silane coupling agents with unmodified HAp unsuccessfully due to the instability of the chemical bond in the presence of water [10, 11].

The surface modification of hydroxyapatite (HAp) with can be very beneficial for the processing of functionally advanced biomaterials with improved mechanical properties and enhanced bone healing and remodeling interactions. There are many reports in the literature describing various agents that have been used to modify the surface of HAp and the factors that motivate research involving surface modification of HAp. HAp has been widely used as an adsorbent for protein purification. Additional factors include improved colloid stability[12], improvement of interfacial bond formation with polymer matrices for both dental[13] and bone applications, enhanced bioactivity, fluorescent probes for imaging, and drug delivery[14,15]. Misra studied the surface modification of HAp via adsorption interactions in aqueous and nonaqueous solvents[16]. Tanaka et al have showed how the native phosphate based chemistry of HAp can be used to improve surface modification interactions[17]. A variety of chemical agents have been used to modify the surface of HAp including molecules from the general classes of natural polysaccharides, silane coupling agents, alcohols, acids, bases, monomers, polymers, and proteins and peptide sequences[18]. The physical properties of HAp can also be modified at the surface by influencing the surface roughness and topography, crystalline order, conformation of surface adsorbed species, and the length scale over which the surface is modified.

The BONE/CRAB Lab has made a contribution to the rich field of literature related to the surface modification of HAp. The purpose of this work is to provide a selected review of the surface modification of HAp with a focus on the research efforts in the BONE/CRAB Lab. The ultimate goal of HAp surface modification studies is to link key aspects of HAp surface modification and lessons learned from the literature with actual surface modification phenomena that natural bone mineral HAp undergoes during the healing and remodeling of bone tissue.

EXPERIMENTAL PROCEDURES
Chemical Reagents

The materials used to synthesize hydroxyapatite were reagent grade calcium nitrate [$Ca(NO_3)_2 \cdot 4H_2O$], ammonium dihydrogen phosphate ($NH_4H_2PO_4$), and ammonium hydroxide (NH_4OH) [Fisher Scientific, Pittsburgh, PA]. Dodecyl alcohol (Sigma-Aldrich, St. Louis, MO) was used for the surface modification reactions. All chemicals were used as received. Deionized water (Millipore Elix water system, Bedford, MA) and anhydrous ethyl alcohol (Pharmco™ Products, Brookfield, CT) were used in the preparation of chemical solutions and particle suspensions.

Synthesis of Nanophase Hydroxyapatite

Nanophase HAp particles were prepared by a chemical precipitation and hydrothermal technique. Stock solutions containing 1.67 molal (m) $Ca(NO_3)_2 \cdot 4H_2O$ and 1.00 m $NH_4H_2PO_4$ were prepared and used for particle synthesis. In a typical reaction, equal masses of both the 1.00

m PO_4^{3-} solution and the 1.67 m Ca^{2+} solution (typically 40 grams) were measured out and the pH of each solution was adjusted to 10 by the addition of NH_4OH. Five ml aliquots of the Ca^{2+} solution were sequentially added to the magnetically stirred PO_4^{3-} solution. The precipitation reaction proceeded according to the following idealized equation

$$10\ Ca(NO_3)_2\cdot 4H_2O + 6\ NH_4H_2PO_4 + 2\ NH_4OH \rightarrow$$
$$Ca_{10}(PO_4)_6(OH)_2 + 8\ NH_4NO_3 + 12\ HNO_3 \qquad (1)$$

The pH of the precipitated HAP slurry was maintained at 10 by the addition of NH_4OH as needed. Precipitation reactions were performed in polypropylene cups. The precipitate slurry was stirred for 24 hr and subsequently aged in a Teflon lined hydrothermal reaction vessel (Parr Instrument Company, Moline, IL) at 130 °C for 6-10 hr. After aging, the particles were washed with deionized water via centrifugation using a Beckman Avanti J-25I Centrifuge (Fullerton, CA). Nanophase HAP was stored in water or ethyl alcohol at specified solids loadings (g/L) until characterization was performed.

Esterification of Nanophase Hydroxyapatite

A suspension of HAp particles in ethyl alcohol was centrifuged and re-dispersed in dodecyl alcohol via shaking and ultrasonication. The resulting HAp-dodecyl alcohol suspensions were heat treated at 115 °C for 24 hours and 190 °C for 3 hr. After aging, the HAP was washed in ethyl alcohol three times before dispersing the HAp particles in ethyl alcohol.

Synthesis of Silica Coated Hydroxyapatite

Dodecyl modified HAP core particles were dispersed in ethyl alcohol at a concentration of 3 g/L. A magnetic stir bar was used to stir the suspension during the coating reaction. Sufficient tetraethylorthosilicate (TEOS) was added to the HAp suspension to vary the silica coating amount from 5 – 75 wt%. The HAp/TEOS suspension in ethanol was allowed to stir for 30 min before the addition of reagent grade ammonium hydroxide in a 1:4 TEOS to NH_4OH volume ratio. Silica coated HAp suspensions were allowed to stir for 24 hr to ensure complete reaction. The silica coated HAp was washed via centrifugation and re-dispersion in fresh solvent. Ethyl alcohol was used for the first wash after centrifugation and removal of the initial supernatant. Subsequent centrifuge cycles (approximately 20) were conducted using water as the solvent at rotational speeds of 2000-9000 rpm for 5 min. The higher centrifuge speeds were required for the higher silica content samples.

PEI modified HAp

50 w/v % aqueous polyethyleneimine solution (Sigma-Aldrich, St. Louis, MO) was diluted with deionized water and adsorbed onto the hydroxyapatite surface. The polyethylene imine has a number average molecular weight of 60,000 and a weight average molecular weight of 750,000. It is a branched chain polymer with a 1:2:1 ratio of primary:secondary:tertiary amine and a branching site at every 3-3.5 nitrogen atoms on the backbone.

Gum Arabic Modified HAp

Gum arabic (TIC Gums, Belcamp, MD) was dissolved in deionized water (Elix, Millipore, Bedford, MA) to prepare a 10 wt% solution. The 10 wt% GA solution was centrifuged at 20,000 rpm (48,384 G) for 15 minutes to remove the coarse fraction of gum arabic, resulting in an amber colored transparent solution. The HAP and gum arabic solutions were mixed with phosphate buffered saline to yield a suspension that contained 1 wt% gum arabic, 1 wt% HAp, 0.15 M NaCl, 0.01 M phosphate buffer at a pH of 7.4. The solutions were ultrasonicated for 1 min prior to observing the settling behavior.

Synthesis of chitosan acetate gels and chitosan modified hydroxyapatite

Chitosan acetate gels (CAG) were made at 1-5 mass% using chitosan, glacial acetic acid, and deionized water. The mass% corresponded to both chitosan and acetic acid concentrations. The CAG gels were prepared by dispersing the chitosan into the aqueous acetic acid solution and allowing the chitosan to dissolve over a 24 hour time period. The CAG-HAp solutions were prepared by mixing 20 g of a 1, 3, and 5 mass% CAG gel with and equal mass (20 g) of an aqueous HAp solution containing 8.9 mass% HAp. The final samples contained 1.77 grams of HAp in 0.5 mass% CAG (HAp[0.5]), 1.5 mass% CAG (HAp[1.5]) and 2.5 mass% CAG (HAp[2.5]). The samples were aged for one month in closed polypropylene cups and washed via centrifugation (14,000 rpm, 10-30 min) a minimum of three times with de-ionized water to remove excess chitosan that was not strongly adsorbed on the HAp particle surface. A portion of the samples was dried for further analysis.

Characterization:

The HAp samples were characterized via FTIR, TGA, CHN analysis, and BET N_2 adsorption. Diffuse reflectance infra-red Fourier transform spectroscopy (DRIFTS) was used to identify and verify the presence of specific functional groups samples. DRIFTS analysis was performed using a Nicolet 500 Series Optical Bench (Madison, WI) with the diffuse reflectance measuring attachment and a KBr detector. Samples were prepared by mixing 10 wt% of the sample with 90 wt% optical spectra grade KBr (Fisher Scientific, Pittsburgh, PA). Pure KBr was used as the reference. Sample spectra were collected over the range of $4000 - 400$ cm^{-1} at a resolution of 4 cm^{-1} and the reported data represents the average of 128-256 scans for a better signal to noise ratio.

Thermal gravimetric analysis (TGA) was performed on the synthesized particles using a thermal analysis system (Shimadzu TGA-50, Kyoto, Japan). Approximately 20-24 mg of chitosan modified HAp powder was ground in a mortar and pestle and placed in an alumina sample pan for TGA analysis. The sample heating rate was 10 °C/min to a maximum temperature of 1000 °C. Carbon, hydrogen, and nitrogen (CHN) elemental analysis for the chitosan modified HAp samples was performed by Prevalare Life Sciences (NY). The specific surface area of the chitosan modified HAp was determined by BET N_2 adsorption using a Quantachrome Nova 1200 Quantasorb (Boyntown, FL).

The colloid stability of HAp with adsorbed PEI was determined via sedimentation studies. The sample size was 30 ml and consisted of 1 g/l HAp in 0.005 M NaNO$_3$. The PEI was added in concentrations of 0.001 to 1 g/l. For electrophoresis studies, 100 ml samples of 0.005 vol% HAp were prepared in 0.005 M NaNO$_3$. PEI was added to the HAp-NaNO$_3$ solution at the specified concentrations of 0.005, 0.05, and 0.5 g/l. For FTIR, 100 ml mixtures of 0.1 g HAp in 0.005 M NaNO$_3$ with PEI at concentrations of 0.0005, 0.05, and 1 g/l were dispersed in an ultrasonic bath. The pH was adjusted with HNO$_3$ or NaOH and mixtures were ultrasonicated again. Samples were then left to sit for 24 hours to reach an equilibrium state.

RESULTS AND DISCUSSION

Silica Modidfied HAp

TEM micrographs of dodecyl alcohol modified HAp and HAp surface modified with 50 wt% silica are shown in Figures 1a and 1b. A silica surface coating can be seen on the HAp particles based on the appearance of rougher surface features on the particles and decreased inter-particle porosity. . In comparison to the uncoated hydroxyapatite (50 m^2/g), silica coatings of 5, 25, 50 and 75 wt% yielded specific surface areas of 55, 93, 70, and 138 m^2/g, respectively. This behavior can be explained based on a hetero-coagulation coating mechanism in which silica clusters of approximately 14 nm in diameter adsorb onto the hydroxyapatite particle surface. The decrease in specific surface area at 50 wt% silica corresponded to the attainment of a complete surface coating. This conclusion was substantiated by the observed resistance of these particles to dissolution in 1 M HCl. However, the acid treatment transformed the silica coated hydroxyapatite core particles to $CaCl_2 \cdot Ca(H_2PO_4)_2 \cdot 2H_2O$ (calcium chloride phosphate hydrate) based on XRD analysis.

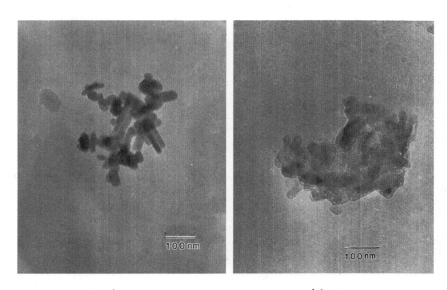

a.)   b.)

Figure 1.) Transmission electron micrograph of a.) dodecyl alcohol treated hydroxyapatite and b.) silica coated HAp (50 wt% silica)

PEI Modified HAp

Table I lists the sedimentation times for 30 ml samples of 1 g/l HAp with increasing PEI concentration. The initial addition of PEI at concentrations below 0.03 g/l resulted in unstable suspensions. A visible observation showed particle flocculation materializing immediately. Large HAP particle agglomerates settled out of solution within 15 minutes. It took 45 more minutes for the remaining HAP agglomerates to completely settle out, leaving a clear supernatant and loosely packed sediment. At PEI concentrations of 0.03 g/l and above, particles remained dispersed after one hour. After 24 hours, the supernatant began to separate into a cloudy phase topped by a clear layer. It took 5 days for particles to completely settle out of suspension. Figure 2 gives a pictorial representation of how the solutions described in Table I appeared after various settling times. Increasing PEI to concentrations greater than 0.03 g/l did not provide any additional benefits to improving particle stability. Sedimentation time remained constant (5 days) for PEI concentrations ranging from 0.03 g/l to 1.0 g/l.

Table I. Sedimentation times of HAp with varying PEI concentrations in solution at pH=7. PEI was added to 30 ml samples of 1g/l HAp.

PEI in solution (g/l)	Sedimentation Time to B* Behavior	Behavior*
0.001	3 hours	After 1 hour: A Fluffy Sediment
0.005	3 hours	After 1 hour: A Fluffy Sediment
0.01	3 hours	After 1 hour: A Fluffy Sediment
0.03	5 days	After 1 hour: C After 24 hours: D Compact Sediment
0.05	5 days	After 1 hour: C After 24 hours: D Compact Sediment
0.1	5 days	After 1 hour: C After 24 hours: D Compact Sediment
0.5	5 days	After 1 hour: C After 24 hours: D Compact Sediment
1.0	5 days	After 1 hour: C After 24 hours: D Compact Sediment

*Codes for settling behavior.
A: Fluffy sediment forms with flocs of particles in supernatant.
B: Particles are completely settled out with clear supernatant.
C: Particles remained completely dispersed.
D. Thin sediment forms with a cloudy supernatant topped with a clear layer.

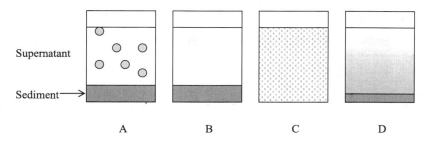

Figure 2 Picture representation of the HAp-PEI suspensions corresponding to Table 1.

The zeta potential vs. pH data for PEI at 0.5 g/l is reported in Figure 3. The iso-electric point of PEI at pH 10.8 is in accord with reports in literature (Hostetler and Swanson, 1974). Plots of the zeta potential vs. pH for PEI adsorbed onto HAp is shown in Figure 2. Regardless of the PEI concentration, the behavior of all particles is similar. The presence of PEI shifts the iso-electric point of the HAP particles to 10.8. The zeta potential data in Figure 3 indicates that PEI stabilizes HAp via an electrostatic mechanism. The zeta potential of the HAp with adsorbed PEI is positive and higher in magnitude between pH 7 and 10.8 than that of HAp. The iso-electric (iep) point shifts from 7 (iep of HAp) to 10.8 (iep of PEI). HAp is stabilized via electrostatic interactions with 0.005 g/l PEI in solution. The additional PEI at 0.05 and 0.5 g/l seems to provide no benefit for electrostatic stabilization. However, sedimentation data (Table I) show that at 0.005 g/l PEI, particles flocculate, despite the enhanced electrostatic interactions. It is not until more PEI is added (concentrations greater than 0.03 g/l) that the particles remain stable. This additional PEI in solution most likely provides a steric hindrance to aggregation.

The steric repulsive forces felt between the PEI adsorbed HAp particles are due to the conformational changes of the PEI molecules with concentration. A study done by Afif et al.[19] investigated the molecular conformation of PEI adsorbed onto silica using electron paramagnetic resonance (EPR) spectroscopy. The labeling is achieved through a reaction between PEI and 3-carboxyl-proxyl, $C_9H_6NO_3$. They found that at low PEI concentrations, the conformation of the chains laid flat on the silica surface. At high concentrations, the conformation depended upon the specific surface area of the silica. For smaller specific surface areas, less space was available for PEI molecules to lay flat. The chains would interact with each, repelling from one another and extending into solution. For higher specific surface areas, there was enough space for PEI chains to lie flat on the surface and maximize contact between charged polymer groups and the silica surface. Accordingly, at low PEI concentration (<0.03 g/l) on HAp, the PEI chains adopt a flat conformation. This explains how PEI can improve the zeta potential of HAp particles without increasing its colloidal stability. There are no extended polymer segments to provide the steric stabilization. At PEI concentrations ≥ 0.03 g/l, the PEI molecules may feel greater repulsive forces between each other, extending their loops and tails away from the particle surface into solution at pH 7. This corresponds to a minimum of 59 mg PEI per m^2 of HAp surface area that is required to prevent HAp aggregation.

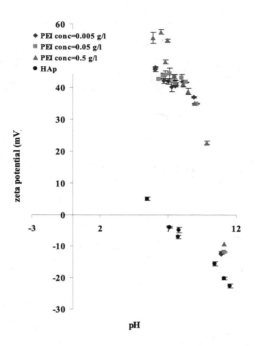

Figure 3 Zeta potential vs. pH for PEI adsorbed on HAp surfaces at PEI concentrations 0.005, 0.05, and 0.5 g/l.

GA modified HAp

Figure 4 shows a photo of a 1 wt% HAp sample dispersed in 1 wt% gum arabic that remained well dispersed after one hour while an identical HAp sample without gum arabic formed a thick HAp sediment layer. The experimental conditions were chosen to correspond to physiological conditions where the colloid stability of HAp is lowest. Gum arabic was just as effective at dispersing HAp in 1 M NaCl while conventional dispersants such as polyacrylic acid were not effective at dispersing HAp in high ionic strength media. The fact that gum arabic effectively disperses HAp in high ionic strength media near its iep indicates that the dispersion mechanism is steric or electrosteric in nature due to charged functional groups on the polymer.

Gum arabic is uniquely tailored for dispersing HAp under physiological conditions. The main attraction of gum arabic involves its similar chemical and structural features with extra cellular matrix proteins (ECMP) such as decorin and osteoadherin which serve important roles in hard tissue biology. Gum arabic exhibits strong interactions with Ca^{2+} and contains approximately 5 wt% calcium as determined by TGA. The water soluble nature of gum arabic makes it a green processing agent (environmentally friendly).

Figure 4 Photo of 1 wt% HAp dispersed in 1 M NaCl aqueous solution with no dispersant (on the right side) and 1 wt% GA (on the left side).

These results have wide reaching potential for the processing and development of HAp/polymer nanocomposites with enhanced properties for bulk HAp components. On a much smaller scale, well dispersed HAp "nano" nanocomposites can also be developed for DNA transfection studies targeted towards treating hard tissue disorders based on the initial work of in vitro gene transfection using a calcium phosphate/DNA complex[20].

CONCLUSIONS AND FUTURE DIRECTIONS

There is a very promising future for studying HAp surface modification. While there are a lot of commercially viable applications that will come from this work, the most exciting results will come from correlating HAp surface modification/adsorption phenomena and in vitro experiments to gain insight into in vivo phenomena and processes in bone healing. Further insights into the relationship between extracellular matrix proteins, and HAp will lead to unique ways of molecular surface engineering of HAp to optimize biological activity such as protein sequence grafting for cell signalling and help in designing BINs that stimulate the correct sequential cascade of events that occur during bone healing for optimum BIN integration into bone tissue[21-23].

REFERENCES

[1]Y. Liu and G. H. Nancollas, Crystallization and colloidal stability of calcium phosphate phases, J. Phys. Chem. B, 101, 3464-3468 (1997).

[2]L. Borum and O. C. Wilson, Jr., Surface Modification of Hydroxyapatite: I Dodecyl Alcohol, Biomaterials, 24(21), 3671-3679 (2003).

[3]O. C. Wilson, Jr. and L. Marshall, Molecularly Dispersed Hydroxyapatite Polymer Nanocomposites, Ceram. Trans., 147, 91-100 (2003).

[4]L. Borum and O. C. Wilson, Jr., "Surface Modification of Hydroxyapatite: II Silica," Biomaterials, 24(21), 3681-3688 (2003).

[5]L. Borum, Enhanced Colloid Stability of Hydroxyapatite, Ph.D. dissertation, University of Maryland, College Park, 2000.

[6]A. C. Roque and O. C. Wilson, Jr., Adsorption of Gum Arabic on Bioceramic Nanoparticles, Mater. Sci. Eng. C, 28, 443-447 (2008).

[7]O. C. Wilson, Jr. and J. Hull, Surface Modification of Nanophase Hydroxyapatite with Chitosan, Mater. Sci. Eng. C, 28, 434-437 (2008).

[8]E. M. Carlisle, Silicon: A possible factor in bone calcification, Science, 167, 279-280 (1970).

[9]L. Hench and H. Paschall, Direct chemical bond of bioactive glass-ceramic materials to bone and muscle, J. Biomed. Mater. Res. 7, 25-42 (1973).

[10]S. Deb, M. Braden, and W. Bonfield, Water absorption characteristics of modified hydroxyapatite bone cements, Biomaterials, 16(14), 1095-1100 (1995).

[11]A. M. Dupraz, J. R. de Wijn, S. A. van der Meer, and K. de Groot, Characterization of silane treated hydroxyapatite powders for use as filler in biodegradable composites, J. Biomed. Mater. Res., 30(2), 231-238 (1996).

[12]P. Somasundaran, J. O. Amankonah, and K. P. Ananthapadmabhan, Mineral—solution equilibria in sparingly soluble mineral systems, Colloids Surf., 15, 309-333 (1985).

[13]D. N. Misra, Adsorption of benzoic acid on pure and cupric ion-modified hydroxyapatite: implications for design of a coupling agent to dental polymer composites. J Dent Res, 65(5), 706-711 (1986).

[14]A. Barroug and M. J. Glimcher, Hydroxyapatite crystals as a local delivery system for cisplatin: adsorption and release of cisplatin in vitro, J. Orthop. Res., 20(2), 274-280 (2002).

[15]A. Barroug, L. T. Kuhn, L. C. Gerstenfeld, and M. J. Glimcher, Interactions of cisplatin with calcium phosphate nanoparticles: in vitro controlled adsorption and release, J. Orthop. Res., 22(4), 703-708 (2004).

[16]D. N. Misra, Adsorption from solutions on synthetic hydroxyapatite: nonaqueous vs. aqueous solvents. J. Biomed. Mater. Res., 48(6), 848-55 (1999).

[17]Tanaka H, Yasukawa A, Kandori K, Ishikawa T. Surface modification of calcium hydroxyapatite with hexyl and decyl phosphates," Colloids Surf A: Physico Eng Aspects 1997;125:53-62.

[18]S. Takemoto, Y. Kusudo, K. Tsuru, S. Hayakawa, A. Osaka, and S. Takashima, Selective protein adsorption and blood compatibility of hydroxyl-carbonate apatites, J. Biomed. Mater. Res., 69A(3), 544-551 (2004).

[19]A. Afif, H. Hommel, A. Legrand, M. Bacquet, E. Gailliez-Degremont, and M. Morcellet, Molecular conformation and mobility of polyamines adsorbed on silica studied by spin labeling, J. Colloid Interface Sci., **211**, 304 (1999).

[20]Graham FL and van der ebb AJ, Virology 1973; 52:456-467.

[21]Hing KA. Bone repair in the twenty-first century: Biology, chemistry or engineering? Phil. Trans. R. Soc. Lond. A, 362:2821-2850 (2004).

[22]O. Wilson, Jr., Bone inspired nanocomposites, in *New Research on Nanocomposites*, Eds. Luis M. Krause et al., (NovaScience Publishers, NY), pp 57-78 (2008).

[23]A. S. Posner, The structure of bone apatite surfaces, J. Biomed. Mater. Res., **19**(3), 241-250 (1985).

Materials for Drug Delivery

NANOPHASE HYDROXYAPATITE IN BIODEGRADABLE POLYMER COMPOSITES AS NOVEL DRUG-CARRYING IMPLANTS FOR TREATING BONE DISEASES AT TARGETED SITES

Huinan Liu and Thomas J. Webster
Division of Engineering, Brown University, Providence, RI, USA.

ABSTRACT

Desirable cytocompatibility properties of nanocrystalline hydroxyapatite (HA) were combined with the tunable degradability and deformability of a select polymer (poly-lactide-co-glycolide, or PLGA) to optimize biological and mechanical properties for bone regeneration. In this study, nano-HA/PLGA composites were used as novel drug delivery systems to treat bone diseases at targeted tissue-implant interfaces. Specifically, a model peptide (DIF-7c) derived from bone morphogenetic protein (BMP-7) was loaded into nano-HA/PLGA composites using a covalent chemical bonding method. Results showed that the model peptide was successfully immobilized onto nano-HA in a greater loading efficiency using amino-silane chemistry. Importantly, a prolonged two-phase peptide release (up to 52 days) was achieved using nano-HA/PLGA drug delivery systems. Ceramic/polymer nanocomposites are promising drug-carrying implant materials for treating bone diseases at targeted sites.

INTRODUCTION

Pharmaceutical agents are often required to stimulate new bone formation for the treatment of bone injuries or diseases (such as bone fracture, osteoporosis and osteosarcoma). However, conventional systemic administrations (such as oral and intravenous administration) of these agents can not effectively reach targeted sites and, thus, they can cause non-specific bone formation in areas not affected by injury or disease. Moreover, even if intentionally delivered or implanted locally to the damaged bone tissue, these agents tend to rapidly diffuse into adjacent tissues due to weak physical bonding to their drug carriers, which limits their potential to promote prolonged bone formation in targeted areas of bone. Therefore, this study explored the chemical method for immobilizing peptides derived from BMPs (bone morphogenetic proteins) to nano-HA (nanophase hydroxyapatite) to promote drug loading efficiency and to achieve controlled release at local disease sites. In addition, the use of nano-HA can increase peptide or protein loading efficiency considering that nano-HA has much larger surface area and much more exposed reaction sites for chemical bonding.

MATERIALS AND METHODS

Synthesis of Nanocrystalline HA

Nanophase HA was synthesized using a wet chemistry precipitation method by mixing solutions of calcium nitrate (Sigma) and ammonium phosphate (Sigma) in an alkaline pH region. Specifically, a 1 M calcium nitrate solution and a 0.6 M ammonium phosphate solution were

185

prepared by dissolving their respective solid state powders in deionized (DI) water separately. The produced ammonium phosphate solution was mixed with DI water which had been adjusted to pH 10 by ammonium hydride. The pre-made 1 M calcium nitrate solution was then added into the mixture of ammonium phosphate and ammonium hydride at a rate of 3.6 ml/min. Precipitation occurred as soon as the calcium nitrate was added.

Precipitation continued for 10 minutes at room temperature with constant stirring. The supernatant was collected, centrifuged to reduce 75% of the solution volume and placed into to a 125 ml Teflon liner (Parr Instrument). The Teflon liner was sealed tightly in a Parr acid digestion bomb 4748 (Parr Instrument) and treated hydrothermally at 200 °C for 20 hours to obtain nanocrystalline HA. The hydrothermal treatment has a great advantage to prepare a stoichiometric, ultrafine HA powder with a homogeneous shape and size distribution due to higher applied pressures than atmospheric. After the hydrothermal treatment, nano-HA particles were rinsed with DI water and dried in an oven at 80 °C for 12 hours.

Design and Synthesis of the Model Peptide DIF-7c

The BMPs have several hundred amino acids, approximately 2~3 nm, depending on the conformation, which are too large and complex to be chemically functionalized onto nanomaterials. These complex secondary structures of the proteins are prone to degradation and as a result, these proteins tend to lose their bioactivity quickly in aqueous physiological conditions. Moreover, short peptides can be attached to drug carriers more efficiently due to their small size. Therefore, it is proposed in this study to deliver short peptides that were derived from bioactive regions of BMP-7 (osteogenic protein-1), instead of the whole BMP proteins, by chemically functionalizing them onto nano-HA.

Chen et al. investigated three short peptides derived from bioactive regions of BMP-7 [1]. These three peptides were composed of 10 amino acids and were designated as peptide a (SNVILKKYRN), b (KPCCAPTQLN) and c (AISVLYFDDS). The results showed that peptide b increased osteoblast proliferation while peptide a and c promoted osteoblast differentiation (e.g. mineralization) [1].

In this study, the peptide c (AISVLYFDDS) was chosen and further modified at its N-terminal with a cysteine-containing spacer to ease chemical conjugation onto the nano-HA particles using amino-silane chemistry followed by a maleimide cross-linker molecule. The peptide with a 12 amino-acid sequence of CKAISVLYFDDS was used as the model peptide and termed as DIF-7c.

The peptide DIF-7c was obtained as carboxyl terminal acids to more than 98.2% purity according to the HPLC profile provided by the manufacturer (GenScript Corporation, USA). The molecular weight of the peptide DIF-7c was 1360.56 g/mol.

Loading Peptide onto Nano-HA/PLGA Composites

As mentioned, the difficulties of drug delivery lie in the efficient loading and controlled release. In this study, a chemical bonding method was used for improving the efficacy of drug delivery.

Preparation of HA_Ps_PLGA Drug Carriers

For chemical bonding, nano-HA was functionalized through amino-silane chemistry under dry conditions to avoid surface contamination and, thus, ensure stability of the peptide [2,3]. First, nano-HA was silanized in 3-aminopropyltriethoxysilane (APTES; Sigma 440140) in anhydrous hexane (Sigma 296090). Second, for substituting a hetero-bifunctional cross-linker for the terminal amine, the silanized nano-HA was coupled with N-succinimidyl-3-maleimido propionate (SMP; also called 3-Maleimidopropionic acid N-hydroxysuccinimide ester, Sigma 358657) in anhydrous N,N-dimethylformamide (DMF; Sigma 494488). Third, the peptide DIF-7c was immobilized onto nano-HA in anhydrous DMF through a reaction between the outer maleimide group with the thiol group of cysteine present in the terminal of DIF-7c. The nano-HA and model peptide conjugates that were bonded using amino-silane chemistry were termed as HA_Ps.

PLGA pellets (50/50 wt.% poly(DL-lactide/glycolide, Polysciences, Inc., Warrington, PA) was dissolved in an organic solvent at 40 °C for 40 minutes. The obtained HA_Ps nanoparticles were then added into PLGA solution. The weight ratio of HA_Ps to PLGA was 30/70. The mixture was sonicated for 10 min at controlled powers to achieve a uniform dispersion of HA_Ps in PLGA. After sonication, the mixture was cast into a Teflon mold, evaporated in air at room temperature for 24 hours, and dried in a vacuum oven at room temperature for 48 hours. These HA_Ps_PLGA films were then cut into 1 cm × 1 cm squares for use in material characterizations and *in vitro* studies.

Preparation of Controls

PLGA with the peptide were used as polymer controls. For this purpose, PLGA was first dissolved in an organic solvent at 40 °C for 40 minutes; and the peptide was added into PLGA solution after PLGA was completely dissolved. The PLGA_peptide mixture was then cast into a Teflon mold, evaporated in air at room temperature for 24 hours, and dried in an air vacuum chamber at room temperature for 48 hours. These PLGA_peptide films were then cut into 1 cm × 1 cm squares and termed as PLGA_P for use in material characterizations and in vitro studies. HA_Ps was also used as ceramic controls.

Characterization of Drug Loading using CBQCA Assay

A novel 3-(4-carboxybenzoyl)quinoline-2-carboxaldehyde (CBQCA, Molecular Probes) fluorescence technique was used to characterize the loading of the peptide onto the nano-HA. This technique could provide ultrasensitive detection of primary amines. Inherently CBQCA is a non-fluorescence molecule, but it becomes highly fluorescent upon reaction with amine groups in the presence of cyanide molecules. CBQCA reacts specifically with primary amines to form conjugates that are highly fluorescent and the sensitivity of detection of CBQCA conjugates could reach the attomole range (10^{-18} moles).

CBQCA reagent solutions were prepared by dissolving the CBQCA (MW = 305.3 g/mol) in dimethylsulfoxide (DMSO, Sigma D2650) (10 mM). Potassium cyanide (KCN, MW = 65.1, Sigma 60178) was dissolved in DI water to give a 10 mM working solution. HA_Ps nanoparticles were exposed to CBQCA and potassium cyanide working solutions for 2 hours at room temperature, and were then visualized under a fluorescence microscope (LEICA DM5500B upright fluoresence microscope). Images were obtained using Image Pro software.

Surface Characterization

The HA_Ps_PLGA drug delivery systems and PLGA_P controls were characterized using a Field Emission Scanning Electron Microscope (FESEM, LEO 1530) at a 3 kV accelerating voltage. The specimens were sputter-coated with a thin layer of gold-palladium, using a Hummer I Sputter Coater (Technics) in a 100 mTorr vacuum argon environment for 3 min at 10 mA of current.

In Vitro Drug Release Profiles

In vitro peptide release kinetics were studied in PBS (pH=7.4). All samples of interest were incubated in PBS under standard cell culture conditions for 52 days. After 1, 3, 5, 7, 30, and 52 days, the supernatants were collected and analyzed. The peptide release from scaffolds into culture solution was determined using a micro-BCA assay (Pierce). Briefly, the peptide DIF-7c standards were prepared by a serial dilution and the working reagent was mixed according to the established protocol. Each standard and unknown samples were aliquoted in 150 μL into a microplate well and mixed thoroughly with the working reagent on a plate shaker for 30 seconds. The reactions were incubated at 37 °C for 2 hours. The microplates were cooled to room temperature and read the absorbance at 562 nm using a spectrophotometer. A standard curve was generated by plotting the average Blank-corrected 562 nm reading for each peptide standard versus its concentration in μg/mL. The peptide concentration in the supernatants was calculated according to the standard curve.

RESULTS AND DISCUSSION

Characterization of Drug Loading

The results of the CBQCA assay demonstrated the success of loading the peptide onto nano-HA chemically, as shown in Figure 1. Nano-HA with chemically loaded peptide produced very good fluorescence (Figure 1d), which indicated the successful attachment of the peptide onto nano-HA. Moreover, in the absence of the CBQCA, APTES treated nano-HA did not fluorescence (image not shown). In contrast, in the presence of CBQCA, APTES treated nano-HA did fluorescence (Figure 1b). Nano-HA after SMP reaction did not fluorescence (Figure 1c), indicating that the amine groups were completely covered by the SMP. The nano-HA (without peptide) did not show fluorescence (Figure 1a), which provided evidence that the CBQCA did not react with HA and only reacted with the amino groups.

(a) Nano-HA:
no fluorescence

(b) Nano-HA after APTES treatment:
fluorescence

(c) Nano-HA after SMP reaction:
no fluorescence

(d) Nano-HA after peptide attachment:
fluorescence

Figure 1: The CBQCA analysis of nano-HA loaded with the model peptide DIF-7c by the chemical bonding method. Fluorescence images are: (a) nano-HA, (b) nano-HA after APTES treatment, (c) nano-HA after SMP reaction, and (d) nano-HA with the chemically attached peptide. Original magnifications are 10x. Scale bars are 500 μm.

Surface Characterization

Scanning electron micrographs suggest that the distribution of nano-HA particles was uniform in the HA_Ps_PLGA drug delivery systems after controlled sonication, even when these HA nanoparticles were chemically functionalized with the peptide DIF-7c (Figure 2a). PLGA_P maintained a very smooth surface similar to the PLGA (Figure 2b).

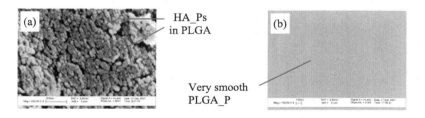

HA_Ps
in PLGA

Very smooth
PLGA_P

Figure 2: SEM images of (a) HA_Ps_PLGA with many nano-features and (b) PLGA_P with very smooth surface. Magnification bars are 200 nm for (a) and 100 nm for (b).

In Vitro Drug Release Profiles

The release of peptide DIF-7c in vitro was studied for up to 52 days, as shown in Figure 3. In Figure 3(a), the single phase drug carriers, including PLGA_P and HA_Ps, all demonstrated one-phase release, although the major release happened at different time points for the HA carrier and the PLGA carrier. Specifically, the HA carrier (HA_Ps) started the peptide release at day 1, while the PLGA carrier did not release any peptide until day 7. At day 30, the HA carrier stopped the peptide release, while the PLGA carrier showed evidence of peptide release. At day 52, the PLGA carrier continuously showed peptide release, while HA carrier did not release any peptide. The HA carrier demonstrated continuous peptide release from day 1 to 7. The total amount of peptide released by the HA_Ps was greater than the PLGA_P during 52 days. It was speculated that the nano-HA had higher peptide loading efficiency compared to the PLGA. That is, chemical fictionalization permitted more peptide to be attached onto nano-HA compared to physical dispersion of peptide in PLGA. In Figure 3(b), the composite drug carrier (HA_Ps_PLGA) demonstrated two-phase release. At phase I (from day 1 to 7), the HA_Ps_PGA demonstrated continuous peptide release at a gradually decreased amount. At phase II, the HA_Ps_PLGA demonstrated increased peptide release from day 30 to 52.

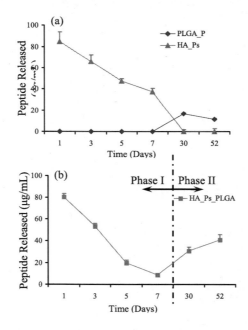

Figure 3: The amount of peptide DIF-7c released from the drug delivery systems of interest to this study. (a) Peptide released from the controls: PLGA_P and HA_Ps. (b) Peptide released from the nanocomposites: HA_Ps_PLGA. Values are mean ± SEM; N=3.

A series of drug therapies are usually necessary after orthopedic surgeries to prevent either infection or inflammation or to induce appropriate natural tissue integration with the implants. Currently, drugs (such as antibiotics, anti-inflammatory drugs and bone growth factors) are typically administered either orally or intravenously. These routes of drug delivery often result in limited bioavailability, thus, requiring high dosages for drugs to be effective at the site of implantation. The ideal situation is delivering drugs directly at the interface of the implant and tissue. In other words, drug carrying scaffolds or implants (such as HA_Ps_PLGA) that are capable of controlling drug release may provide a promising approach for treating bone diseases at targeted sites.

CONCLUSIONS

Results of this study demonstrated three different drug release profiles achieved by using various drug carriers. The drug loading efficiency are related to the drug carriers and the loading methods. Single phase drug carriers (such as HA_Ps and PLGA_P) provided one-phase release profiles. The nanocomposite drug carrier (such as HA_Ps_PLGA) demonstrated a two-phase release profile. Importantly, a prolonged peptide release (up to 52 days) was achieved on the HA_Ps_PLGA drug delivery systems. The drug carriers and the drug loading methods are very important factors that should be considered when designing the next generation of drug carrying orthopedic prostheses for various clinical applications. The appropriate drug carriers and drug loading methods should be carefully chosen for specific applications. This study presented a useful guideline for designing more effective, controlled drug delivery systems according to requirements of specific applications.

ACKNOWLEDGMENT

The authors would like to thank the NSF for a Nanoscale Exploratory Research (NER) grant.

REFERENCES

[1] Chen Y, Webster TJ. Simple structure, easily functionalized and controlled release bioactive BMP-7 short peptides for orthopaedic applications. Journal of Oral Implantology, in press.

[2] Hong HG; Jiang M; Sligar SG; Bohn PW. Cysteine-specific surface tethering of genetically engineered cytochromes for fabrication of metalloprotein nanostructures. Langmuir. 10(1): 153-158; 1994.

[3] Balasundaram G, Sato M, Webster TJ. Using hydroxyapatite nanoparticles and decreased crystallinity to promote osteoblast adhesion similar to functionalizing with RGD. Biomaterials. 27(14): 2798-2805; 2006.

Author Index